International
REVIEW OF
Neurobiology

Volume 96

International REVIEW OF Neurobiology

Volume 96

Pharmacology of 5-HT6 Receptors, Part II

EDITED BY

FRANCO BORSINI

Sigma-Tau Industrie Farmaceutiche
Riunite S.P.A.,
Pomezia, Italy

ELSEVIER

AMSTERDAM • BOSTON • HEIDELBERG • LONDON
NEW YORK • OXFORD • PARIS • SAN DIEGO
SAN FRANCISCO • SINGAPORE • SYDNEY • TOKYO
Academic Press is an imprint of Elsevier

Academic Press is an imprint of Elsevier
360 Park Avenue South, New York, NY 10010-1700
525 B Street, Suite 1900, San Diego, California 92101-4495, USA
32 Jamestown Road, London NW1 7BY, UK

This book is printed on acid-free paper.

For information on all Elsevier publications
visit our website at books.elsevier.com

ISBN: 978-0-12-385902-0

PRINTED AND BOUND IN THE UNITED STATES OF AMERICA
11 12 13 14 10 9 8 7 6 5 4 3 2 1

CONTENTS

The 5-HT_6 Receptor as a Target for Developing Novel Antiobesity Drugs

David Heal, Jane Gosden and Sharon Smith

Behavioral and Neurochemical Pharmacology of 5-HT_6 Receptors Related to Reward and Reinforcement

Gaetano Di Chiara, Valentina Valentini and Sandro Fenu

5-HT_6 Receptor Ligands and their Antipsychotic Potential

Jørn Arnt and Christina Kurre Olsen

5-HT_6 Receptor Ligands as Antidementia Drugs

ELLEN SIOBHAN MITCHELL

Other 5-HT_6 Receptor-Mediated Effects

FRANCO BORSINI

CONTRIBUTORS

Numbers in parentheses indicate the pages on which the authors contributions begin.

Jørn Arnt (141), Lundbeck Research Denmark, H Lundbeck A/S, DK-2500 Valby, Denmark

Franco Borsini (189), Sigma-Tau Industrie Farmaceutiche, Riunite S.P.A., Pomezia, Italy

C. Castillo (27), Escuela Superior de Medicina del IPN, México

Lee A. Dawson (1), Neurosciences Product Creation Unit, Eisai Limited Hatfield, Hertfordshire AL10 9SN, UK

Gaetano Di Chiara (111), Department of Toxicology, University of Cagliari, 09124 Cagliari, Italy

Sandro Fenu (111), Department of Toxicology, University of Cagliari, 09124 Cagliari, Italy

Jane Gosden (73), RenaSci Consultancy Limited, BioCity, Nottingham NG1 1GF, UK

David Heal (73), RenaSci Consultancy Limited, BioCity, Nottingham NG1 1GF, UK

Magdalena Jastrzębska-Więsek (49), Department of Clinical Pharmacy, Jagiellonian University Medical College, 30-688 Krakow, Poland

Alfredo Meneses (27), Department of Pharmacobiology, CINVESTAV, México City 14330, México

Ellen Siobhan Mitchell (163), Unilever Research and Development, Vlaardingen 3135 XB, The Netherlands

Christina Kurre Olsen (141), Lundbeck Research Denmark, H Lundbeck A/S, DK-2500 Valby, Denmark

G. Pérez-García (27), Department of Pharmacobiology, CINVESTAV, México City 14330, México

T. Ponce-Lopez (27), Department of Pharmacobiology, CINVESTAV, México City 14330, México

Sharon Smith (73), RenaSci Consultancy Limited, BioCity, Nottingham NG1 1GF, UK

R. Tellez (27), Department of Pharmacobiology, CINVESTAV, México City 14330, México

Valentina Valentini (111), Department of Toxicology, University of Cagliari, 09124 Cagliari, Italy

Anna Wesołowska (49), Department of Clinical Pharmacy, Jagiellonian University Medical College, 30-688 Krakow, Poland

PREFACE

Ligands to 5-HT_6 receptors represent so far an unresolved mystery. Agonists and antagonists are defined by *in vitro* techniques for receptor binding and second messenger systems. However, both agonists and antagonists have been claimed to exert similar *in vivo* pharmacological effects, even if information on metabolic pattern is missing. This may explain why 5-HT_6 receptor antagonism may induce (Bentley *et al.*, 1999; Bos *et al.*, 2001; Bourson *et al.*, 1995; Lindner *et al.*, 2003; Marcos *et al.*, 2008b; Reavill and Rogers, 2001; Sleight *et al.*, 1998) or may not induce (Hamon *et al.*, 1999; Otano *et al.*, 1999; Russell and Dias, 2002; Stean *et al.*, 2002; Yoshioka *et al.*, 1998) the so-called behavioral syndrome.

Furthermore, pharmacokinetics often does not correlate with pharmacodynamics. For example, (1) the antagonist SB-357134 had its brain and plasma T_{max} after 1 h but its anticonvulsant effects peaked at 4–6 h (Stean *et al.*, 2002); (2) the anticonvulsant effects induced by oral SB-271046 appeared in less than 30 min, when blood and brain levels are not yet appreciable (Routledge *et al.*, 2000); (3) the maximum brain and plasma concentrations of oral SB-399885 were between 3 and 4 h, respectively, but its best effects on extracellular acetylcholine in the cortex were after 1 h and its anticonvulsant effects after 6 h (Hirst *et al.*, 2006); and (4) the peak effect of compound 11 on acetylcholine extracellular levels was after 20 min, despite its plasmatic T_{max} was at 2 h (Riemer *et al.*, 2003).

In line with these inconsistencies, both 5-HT_6 receptor antagonists and agonists have been reported in animal models to possess antidepressant potential (Chapter 3 by Wesołowska and Jastrzbska-Wisek), to be cognitive enhancers (Chapter 6 by Arnt and Olsen, Chapter 2 by Meneses, Chapter 7 by Mitchell), to exert antiobesity effects (Chapter 4 by Heal *et al.*). The uncertainty of the mechanism of action of 5-HT_6 receptor ligands is also revealed by brain microdialysis studies (Chapter 1 by Dawson, Chapter 5 by Di Chiara *et al.*).

However, it should be considered that the expression of 5-HT_6 receptors may depend on circulating adrenal corticoids (Holmes, 2008; Marcos *et al.*, 2008a; Yau *et al.*, 1997). Thus, the level of stress of animals may influence the *in vivo* effects of 5-HT_6 receptor ligands. Another aspect to be considered is that 5-HT_6 receptors undergo rapid desensitization (Max *et al.*, 1995). Thus, some activities of 5-HT_6 agonists may depend on their agonistic properties at shorter times, but, due to rapid receptor desensitization, an antagonistic action may appear at longer times. If so, stress, time course of the effects, and time of *in vitro* incubation may become crucial to understand some discrepancies. Consequently, at present, no one can affirm with certainty that a particular pharmacological effect is due to agonistic or antagonistic properties of 5-HT$_6$ receptor ligands. This is also reflected by the decreased interest of pharmaceutical companies in the 5-HT$_6$ field.

It is also difficult to have a clear picture of 5-HT$_6$ receptor distribution in the brain. The distribution of 5-HT$_6$ receptors may depend on the radiolabeled compound or by the antibody, or by the use of anesthetics. In fact, receptor density in cloned cells ranges between 1.6 and 6.1 pmol/mg, if [^{3}H]Ro63-0563 (Boess *et al.*, 1998), [^{125}I]SB-258585 (Hirst *et al.*, 2000), [^{3}H]5-HT (Boess *et al.*, 1997), or [^{3}H]LSD (Hirst *et al.*, 2000) is used. This is an intriguing finding because receptor saturation analysis cannot be explained by the presence of multiple binding sites (Boess *et al.*, 1998; Hirst *et al.*, 2000). Another important aspect is the difference in 5-HT$_6$ receptor density between native tissues and recombinant cells, the density being about 30 times lower in native tissues than in cloned cells (Hirst *et al.*, 2000). Several researchers have reported the presence of 5-HT$_6$ receptors in brain regions such as striatum, nucleus accumbens, hippocampus, or olfactory tubercle (Gerard *et al.*, 1997; Hamon *et al.*, 1999; Hirst *et al.*, 2000, 2003; Roberts *et al.*, 2002; Yoshioka *et al.*, 1998). However, the distribution of 5-HT$_6$ receptors in rats does not find a general consensus for certain regions, such as hypothalamus and globus pallidus (Gerard *et al.*, 1997; Hamon *et al.*, 1999). There are few mismatches between the localization of the 5-HT$_6$ receptor and its mRNA (Gerard *et al.*, 1996; Hamon *et al.*, 1999; Ward and Dorsa, 1996). For example, intense 5-HT$_6$ receptor density was observed in cerebellum, where only weak to moderate levels of 5-HT$_6$ receptor mRNA were found. Another mismatch is in hippocampus, where a few 5-HT$_6$ receptors were seen and high levels of 5-HT$_6$-mRNA were detected. Thus, it seems that the 5-HT$_6$ receptor in those regions is formed in somata and then moves to dendrites or axons. 5-HT$_6$ receptors, which appear in the brain from the 12th day of fetal life, are present on GABAergic cells (Ward and Dorsa, 1996), but not on serotonergic (Gerard *et al.*, 1996) and dopaminergic neurons (Hamon *et al.*, 1999). The presence or absence of 5-HT$_6$ receptors on cholinergic neurons depends on the technique used: electrophysiology (Tassone *et al.*, 2010) or biochemistry (Woolley *et al.*, 2001), respectively.

This book wants to evidence the state of the art in the field of 5-HT$_6$ receptor *in vivo* physiology and pharmacology. We hope that this book may represent reference information for all researchers who work in this field.

References

Bentley, J.C., Bourson, A., Boess, F.G., Fone, K.C.F., Marsden, C.A., Petit, N., and Sleight, A.J. (1999). Investigation of stretching behaviour induced by the selective 5-HT$_6$ receptor antagonist, Ro04-6790, in rats. *Br. J. Pharmacol.* **126**, 1537–1542.

Boess, F.G., Monsma Jr., F.J., Carolo, C., Meyer, V., Rudler, A., Zwingelstein, C., and Sleight, A.J. (1997). Functional and radioligand binding characterization of rat 5-HT_6 receptors stably expressed in HEK293 cells. *Neuropharmacology* **36**, 713–720.

Boess, F.G., Riemer, C., Bos, M., Bentley, J., Bourson, A., and Sleight, A.J. (1998). The 5-hydroxytryptamine$_6$ receptor-selective radioligand [^{3}H]Ro63-0563 labels 5-hydroxytryptamine receptor binding sites in rat and porcine striatum. *Mol. Pharmacol.* **54**, 577–583.

Bos, M., Sleight, A.J., Godel, T., Martin, J.R., Riemer, C., and Stadler, H. (2001). 5-HT_6 receptor antagonists: lead optimisation and biological evaluation of *N*-aryl and *N*-heteroaryl 4-amino-benzene sulfonamides. *Eur. J. Med. Chem.* **36**, 165–178.

Bourson, A., Borroni, E., Austin, R.H., Monsma, F.J., and Sleight, A.J. (1995). Determination of the role of the 5-HT_6 receptor in the rat brain: a study using antisense oligonucleotides. *J. Pharmacol. Exp. Ther.* **274**, 173–180.

Codony, X., Burgueno, J.K., Ramirez, M.J., and Vela, M. (2010). Signal transduction: second messenger systems. *Int. Rev. Neurobiol.* **64**, 89–110.

Gerard, C., El Mestkawy, S., Lebrand, C., Adrien, J., Ruat, M., Traiffort, E., Hamon, M., and Martres, M.-P. (1996). Quantitative RT-PCR distribution of serotonin 5-HT_6 receptor mRNA in the central nervous system of control or 5,7-dihydroxytryptramine-trated rats. *Synapse* **23**, 164–173.

Gerard, C., Martres, M.-P., Lefevre, K., Miquel, M.-C., Vergé, D., Lanfumey, L., Doucet, E., Hamon, M., and El Mestkawy, S. (1997). Immuno-localization of serotonin in 5-HT_6 receptor-like material in the rat central nervous system. *Brain Res.* **746**, 207–219.

Hamon, M., Doucet, E., Lefevre, K., Miquel, M.-C., Lanfumey, L., Insausti, R., Frechilla, D., Del Rio, J., and Vergé, D. (1999). Antibodies and antisense oligonucleotide for probing the distribution and putative functions of central 5-HT_6 receptors. *Neuropsychopharmacology* **21**, 68S–76S.

Hirst, W.D., Abrahamsen, B., Blaney, F.E., Calver, A.R., Aloj, L., Price, G.W., and Medhurst, A.D. (2003). Differences in the central nervous system distribution and pharmacology of the mouse 5-hydroxytryptamine-6 receptor compared with rat and human receptors investigated by radioligand binding, site-directed mutagenesis, and molecular modelling. *Mol. Pharmacol.* **64**, 1295–1308.

Hirst, W.D., Minton, J.A.L., Bromidge, S.M., Moss, S.F., Latter, A.J., Riley, G., Routledge, C., Middlemiss, D.N., and Price, G.W. (2000). Characterization of [^{125}I]-SB-258585 binding to human recombinant and native 5-HT_6 receptors in rat, pig, and human brain tissue. *Br. J. Pharmacol.* **130**, 1597–1605.

Hirst, W.D., Stean, T.O., Rogers, D.C., Sunter, D., Pugh, P., Moss, S.F., Bromidge, S.M., Riley, G., Smith, D.R., Bartlett, S., Heidbreder, C.A., Atkins, A.R., Lacroix, L.P., Dawson, L.A., Foley, A.G., Regan, C.M., and Upton, N. (2006). SB-399885 is a potent, selective 5-HT_6 receptor antagonist with cognitive enhancing properties in aged rat water maze and novel recognition models. *Eur. J. Pharmacol.* **553**, 109–119.

Holmes, A. (2008). Genetic variation in cortico-amygdala serotonin function and risk for stress-related disease. *Neurosci. Biobehav. Rev.* **32**, 1293–1314.

Lindner, M.D., Hodges, D.B., Hogan, J.B., Orie, A.F., Corsa, J.A., Barten, D.M., Olson, C., Robertson, B.J., Guss, V.L., Gillman, K.W., Starrett Jr., J.E., and Gribkoff, V.K. (2003). An assessment of the effects of serotonin 6 (5-HT_6) receptor antagonists in rodent models of learning. *J. Pharmacol. Exp. Ther.* **307**, 682–691.

Marcos, B., Aisa, B., and Ramirez, M.J. (2008a). Functional interaction between 5-HT6 receptors and hypothalamic-pituitary-adrenal axis: cognitive implications. *Neuropharmacology* **54**, 708–714.

Marcos, B., Chuang, T.T., and Ramirez, M.J. (2008b). Effects of 5-HT_6 receptor antagonism and cholinesterase inhibition in models of cognitive impairment in the rat. *Br. J. Pharmacol.* **155**, 434–440.

Max, S.I., Monsma, F.J., and Sibley, D.R. (1995). Agonist-induced desensitization of 5-HT_6 serotonin receptor-coupled adenylyl cyclase in stably transfected HEK-293 cells. *J. Serotonin Res.* **2**, 101–116.

Otano, A., Frechilla, D., Cobreros, A., Cruz-Orive, L.M., Insausti, A., Insausti, R., Hamon, M., and Del Rio, J. (1999). Anxiogenic-like effects and reduced stereological counting of immunolabelled 5-hydroxytryptamine$_6$ receptors in rat nucleus accumbens by antisense oligonucleotides. *Neuroscience* **92**, 1001–1009.

Reavill, C., and Rogers, D.C. (2001). The therapeutic potential of 5-HT$_6$ receptor antagonists. *Curr. Opin. Investig. Drugs* **2**, 104–109.

Riemer, C., Borroni, E., Levet-Trafit, B., Martin, J.R., Poli, S., Porter, R.H.P., and Bos, M. (2003). Influence of the 5-HT$_6$ receptor on acetylcholine release in the cortex: pharmacological characterization of 4-(2-bromo-6-pyrrolidin-1-ylpyridine-4-sulfonyl)phenylamine, a potent and selective 5-HT$_6$ receptor antagonist. *J. Med. Chem.* **46**, 1273–1276.

Roberts, J.., Reavill, C., East, S.Z., Harrison, P.J., Patel, S., Routledge, C., and Leslie, R.A. (2002). The distribution of 5-HT$_6$ receptors in rat brain: an autoradiographic binding study using the radiolabelled 5-HT$_6$ receptor antagonist [^{125}I]SB-258585. *Brain Res.* **934**, 49–57.

Routledge, C., Bromidge, S.M., Moss, S.F., Price, G.W., Hirst, W., Newmann, H., Riley, G., Gager, T., Stean, T., Upton, N., Clarke, S.E., Brown, A.M., and Middlemiss, D.N. (2000). Characterization of SB-271046: a potent, selective and orally active 5-HT$_6$ receptor antagonist. *Br. J. Pharmacol.* **130**, 1606–1612.

Russell, M., and Dias, R. (2002). Memories are made of this (perhaps); a review of serotonin 5-HT$_6$ receptor ligand and their biological functions. *Curr. Top. Med. Chem.* **2**, 643–654.

Sleight, A.J., Boess, F.G., Bos, M., Levet-Trafit, B., Riemer, C., and Bourson, A. (1998). Characterization of Ro 04-6790 and Ro 63-0563: potent and selective antagonists at human and rat 5-HT$_6$ receptors. *Br. J. Pharmacol.* **1245**, 556–562.

Stean, T.O., Hirst, W.D., Thomas, D.R., Price, G.W., Rogers, D., Riley, G., Bromidge, S.M., Serafinowska, H.T., Smith, D.R., Bartlett, S., Deeks, N., Duxon, M., and Upton, N. (2002). Pharmacological profile of SB-357134: a potent, selective, brain penetrant, and orally active 5-HT$_6$ receptor antagonist. *Pharmacol. Biochem. Behav.* **71**, 645–654.

Tassone, A., Madeo, G., Sciamanna, G., and Pisani, A. (2010). Electrophysiology of 5-HT$_6$ receptors. *Int. Rev. Neurobiol.* **94**, 111–128.

Ward, R., and Dorsa, D.M. (1996). Colocalization of serotonin receptor subtypes 5-HT$_{2A}$, 5-HT$_{2C}$, and 5-HT$_6$ with neuropeptides in rat striatum. *J. Comp. Neurol.* **370**, 405–414.

Woolley, M.L., Bentley, J.C., Sleight, A.J., Marsden, C.A., and Fone, K.C.F. (2001). A role for 5-HT$_6$ receptors in retention of spatial learning in the Morris water maze. *Neuropharmacology* **41**, 210–219.

Yau, J.L., Noble, J., Widdowson, J., and Seckl, J.R. (1997). Impact of adrenalectomy on 5-HT$_6$ and 5-HT$_7$ receptor gene expression in the rat hippocampus. *Brain Res. Mol. Brain Res.* **45**, 182–186.

Yoshioka, M., Matsumoto, M., Togashi, H., Mori, K., and Saito, H. (1998). Central distribution and function of 5-HT$_6$ receptor subtype in the rat brain. *Life Sci.* **62**, 1473–7147.

THE CENTRAL ROLE OF 5-HT$_6$ RECEPTORS IN MODULATING BRAIN NEUROCHEMISTRY

Lee A. Dawson

Neurosciences Product Creation Unit, Eisai Limited, Hatfield, Hertfordshire, AL10 9SN, UK

I. Introduction

The 5-HT$_6$ receptor is one of the most recently discovered of the serotonin superfamily of receptors. It was discovered by two independent groups using molecular cloning technologies, which were at their peak back in the early 1990s, and was isolated from rat striatum (Monsma *et al.*, 1993; Ruat *et al.*, 1993). The human homolog was subsequently discovered by Kohen *et al.* (1996). The 5-HT$_6$ receptor is a part of the seven-transmembrane G-protein-coupled receptor family that positively couples to Gαs subunits, and as such stimulates adenylate cyclase and increases cAMP signaling cascades. *In situ* hybridization and Northern blot analysis revealed that 5-HT$_6$ receptors are expressed almost exclusively within the mammalian central nervous system (CNS) (Monsma *et al.*, 1993; Ruat *et al.*, 1993; Ward *et al.*, 1995). Regional distribution studies revealed that the highest levels of expression are found in olfactory tubercle, striatal areas, cerebral cortex, and subfields of the hippocampus (Gerard *et al.*, 1996, 1997; Hamon *et al.*, 1999; Monsma *et al.*, 1993; Roberts *et al.*, 2002; Ruat *et al.*, 1993). These studies also suggested that 5-HT$_6$ receptors are largely expressed within the projection regions and not in cell

INTERNATIONAL REVIEW OF NEUROBIOLOGY, VOL. 96
DOI: 10.1016/B978-0-12-385902-0.00001-2

bodies of the raphe nuclei, suggesting a postsynaptic role for the receptor. This localization pattern, together with the early observation that several antidepressants (e.g., amoxapine, amitriptyline, and mianserin) and many of the atypical antipsychotics (e.g., clozapine, olanzapine, and sertindole) show high to moderate affinity (Roth *et al.*, 1994) for the 5-HT_6 receptor, has led to the suggestion that it may have therapeutic utility in the treatment of various neurological disorders.

The chronological history of this receptor has been reviewed in depth in a previous chapter, but elucidating the exact role of this receptor and its neurophysiological role in modulating brain processes has been a slow and steadily evolving story, despite being discovered over 15 years ago. The early studies in preclinical species utilized antisense oligonucleotide administration and behavioral observations in an effort to understand the functional role of 5-HT_6 receptors (Bourson *et al.*, 1995). These studies demonstrated that intracerebroventricular (i.c.v.) administration of antisense oligonucleotide, to disrupt 5-HT_6 receptor-mediated signaling, produced a syndrome of behaviors that included yawning, chewing, and stretching. These behaviors were attenuated by the cholinergic receptor antagonist atropine, thus indicating that reducing the activity of the 5-HT_6 receptor had a modulatory role on cholinergic neurotransmission (Bourson *et al.*, 1995). Later studies by the same group, again using i.c.v. administration of antisense oligonucleotide, showed that cognitive performance in a rat water maze paradigm could be improved (Bentley *et al.*, 1999; Woolley *et al.*, 2001), presumably mediated via the enhanced cholinergic function implied from previous studies. Interestingly, there were also effects on the feeding behavior of these animals (Bentley *et al.*, 1999; Woolley *et al.*, 2001; the neurobiology of these observations and therapeutic implication will be covered in a later chapter). Taken together, these early data suggested that the 5-HT_6 receptor may play a modulatory role on the cholinergic system. Following these antisense oligonucleotide studies, a number of selective 5-HT_6 receptor small-molecule antagonists were discovered. Subsequently, a steady flow of improved antagonist (i.e. showing enhanced DMPK and CNS penetration), and more recently, small-molecule agonist ligands began to appear in the literature. These were quickly used to more fully elucidate the role of the 5-HT_6 receptor in modulating brain neurochemistry and the efficacy that resulted (at least preclinically). This chapter will describe the use of these ligands, using preclinical sampling techniques such as *in vivo* microdialysis, to demonstrate the modulatory role of the 5-HT_6 receptor on multiple neurotransmitter systems and networks within mammalian brain. The underlying neurobiology of the observed changes will also be discussed along with the potential therapeutic implications and utility of both agonist and antagonist ligands in a range of neurobiological disorders.

II. Effects of 5-HT_6 Receptor Antagonist and Agonist Ligands

A. Cholinergic Neurotransmission

As highlighted above early studies using antisense oligonucleotide approaches implicated a role for the 5-HT_6 receptor in the modulation of cholinergic function within the rodent brain. With the development of the early selective small-molecule 5-HT_6 receptor antagonist ligands, the functional correlate for a modulatory role on cholinergic neurotransmission was further strengthened. In rats habituated to observation cages, Ro 04-6790 produced a behavioral syndrome similar to that seen following treatment with antisense oligonucleotides designed to reduce the expression of 5-HT_6 receptors. This behavioral syndrome consisted of stretching, yawning, and chewing (Sleight *et al.*, 1998). Bourson *et al.* (1998) and Bentley *et al.* (1999) similarly demonstrated that Ro 04-6790 induced stretching behavior and also inhibited 6-OHDPAT lesion-induced rotational behavior. Both phenomena were significantly attenuated by the centrally active muscarinic antagonists scopolamine and atropine but not by the *N*-methylscopolamine or by the dopamine D_2 receptor antagonist haloperidol. Similarly, Routledge *et al.* (2000) showed that an alternative 5-HT_6 receptor antagonist, SB-271046 (Bromidge *et al.*, 1999) potentiated the acetylcholine esterase inhibitor physostigmine-induced chewing behaviors. This occurred at similar dose ranges as those that were able to enhance cognitive performance (Hatcher *et al.*, 2005; Rogers and Hagan, 2001; Rogers *et al.*, 1999) and reverse cholinergic-induced impairments (Foley *et al.*, 2004; Liebnen *et al.*, 2005) in various models of learning and memory.

Despite all of these early data implicating a direct role for 5-HT_6 receptors on cholinergic neurotransmission, no direct neurochemical evidence was reported for quite some time. Sleight *et al.* (1999) presented an abstract communication that showed that Ro 65-7199 was able to enhance acetylcholine (ACh) output in both the cortex and the hippocampus of rats using *in vivo* microdialysis. This occurred at doses that had been shown to be efficacious in memory models (Sleight *et al.*, 1999). This same group later demonstrated that a further molecule from their chemical series (*N*-(pyridin-4-yl)-4-amino-benzene sulfonyl phenylamine; compound **11**) also produced an increase in frontal cortex ACh levels, reaching a maximum of 120% above preinjection baseline following a 30 mg/kg oral dosing (Riemer *et al.*, 2003). The temporal profile of this increase showed a maximum response at 20–40 min postdose, which then steadily declined. The PK profile of this molecule largely followed this temporal profile of neurochemical effects, although the maximum concentration was achieved at 1 h postdose and had a half-life of 4 h, which far surpassed the neurochemical readout. Subsequently, two further 5-HT_6 receptor antagonists, SB-271046 and SB-399885, have been shown to enhance forebrain ACh levels in microdialysis studies in the rat. SB-271046 (10 mg/kg s.c.)

produced a 150% increase in extracellular levels of ACh in the rat hippocampus (Zhang *et al.*, 2007). Maximum increases were observed at 40 min post administration and were then sustained for the duration of the study, that is, 160 min postdose (Zhang *et al.*, 2007). The follow-on molecule from SB-271046, SB-399885 (which shows improved DMPK characteristics and, in particular, enhanced oral bioavailability and brain penetration Hirst *et al.* (2006)) produced a dose-related increase in extracellular levels of ACh in the medial prefrontal cortex (mPFC) of the rat. Maximum increases of 200% (at 10 mg/kg p.o.) were achieved at 1 h post dose and were maintained for the duration of the study (i.e. 4 h post dose). These doses correlated well with those that were shown to be efficacious in the rat novel object recognition (NOR) paradigm of recognition memory (Hirst *et al.*, 2006). Interestingly, concurrent *ex vivo* receptor occupancy studies revealed that striatal receptor occupancy was still >80% at 4 h, which was the time required to attain the maximum brain concentration of this molecule (Hirst *et al.*, 2006). Taken together one may conclude that a high level of occupancy, with this competitive receptor antagonist, was required to enhance extracellular levels of ACh and concurrently produce a cognitive enhancement. Perhaps the most recently described 5-HT_6 receptor antagonist is SB-742457. This molecule is of particular interest because it is one of the four molecules that have gone into a clinical population (together with PRX-07034, SYN-114, SUVN502, and more recently SAM531; Table I) and was the first example within the mechanistic class to show cognitive enhancing efficacy in Alzheimer's patients (Upton *et al.*, 2008). In *in vivo* microdialysis studies (Chuang *et al.*, 2006) SB-742457 produced an increase in extracellular levels of ACh in the rat mPFC. A maximum two fold increase was observed with 10 mg/kg p.o. 30–45 min post administration and the increase was maintained for the duration of the sampling period (265 min). Interestingly, no minimal effective dose was identified in this study with a similar two fold increase being observed at 1 and 3 mg/kg p.o., suggesting that this molecule has an increased *in vivo* potency versus those compounds that preceded it.

It would therefore appear that a range of structurally diverse 5-HT_6 receptor antagonist molecules have now been shown to enhance extracellular levels of ACh. These molecules, irrespective of dose or route administered, have all produced an approximately two fold increase in extracellular ACh levels at time points and durations of efficacy commensurate with their PK profile. The reason for this ceiling effect is likely that the tonic influence that serotonin has on hippocampal and cortical cholinergic output is finite. It has been demonstrated that the 5-HT_6 receptor is not localized directly on cholinergic neurons (Marcos *et al.*, 2006; Woolley *et al.*, 2000, 2004). Simplistically, it may seem somewhat paradoxical that a $G\alpha s$ coupled receptor, when antagonized, would actually increase neurotransmitter output. Localization data have suggested that the 5-HT_6 receptors are colocalized with markers of GABAergic function (Gerard *et al.*, 1996, 1997; Hamon *et al.*, 1999; Woolley *et al.*, 2000) and hence play a role in modulating

Table I
5-HT$_6$ RECEPTOR ANTAGONISTS IN CLINICAL DEVELOPMENT: CURRENT STATUS.

Compound (Company)	Study; Indication	Clinical Data and Status
SB-742457 (GSK)	Phase II; AD	In several phase I studies, SB-742457 was found to be well tolerated, with a safety profile similar to that of placebo. Terminal half-life was >24 h; at a dose of 35 mg, 5-HT$_6$ receptor occupancy in the brain was >80%. Two phase II trials have now been completed. Preliminary data suggested efficacy in comparison to aricept (reported at CINP May 2008). Adjunctive study with aricept/donepezil and SB-742457 (15 and 35 mg) is ongoing
SAM-531 (PF-5212365; WAY-262531) (Wyeth)	Phase II; AD	A double-blind phase IIa study included 74 patients with mild to moderate AD who were randomized to 4 weeks of treatment with SAM-531 0.5, 1.5, 3.0, or 5.0 mg o.d. or placebo. SAM-531 appeared to be safe, with most adverse events mild and none serious. PK parameters included t_{max} values of 1.5–4 h, C_{max} values of 15–192 ng/mL, and AUC (0–24 h) of 268–3187 ng h/mL. A trend towards efficacy was seen with SAM-531 (1.5, 3.0, and 5.0 mg) in terms of improvement from baseline on Alzheimer's Disease Assessment Scale-Cognitive Subscale assessments. Patients given the three highest doses of SAM-531 also tended to make fewer errors at week 4 on the CANTAB Paired Associates Learning test than those given placebo. Mini-Mental State Examination results were not indicative of a clear effect of treatment (reported at ICAD July 2010). A phase II study is underway examining three doses of SAM531 (1.5, 3, and 5 mg) versus donepezil
SGS-518 (Lu AE58054) (Lundbeck/ Lilly)	Phase II; schizophrenia and AD	In phase I studies, SGS-518 was well tolerated in both a dose-ranging and a multidose cohort. Encouragingly, in a small trial involving 20 schizophrenia patients stable on antipsychotic medication SGS-518 produced a dose-proportionate improvement in cognition as determined using the Brief Assessment of Cognition in Schizophrenia scale. This effect reached significance at the highest dose tested (240 mg), and no dose-limiting adverse effects were apparent. However, the schizophrenia studies were subsequently terminated due to lack of efficacy. A phase II study adjunctive to donepezil in AD patients is ongoing
PRX-07034 (EPIX)	Phase I; AD and obesity	A number of phase I trials have been completed with PRX-07034. Most significant has been a multiple ascending-dose study in which PRX-07034 was administered once daily for 28 days to 33 obese, but otherwise healthy, adults. The compound was generally well tolerated at up to 600 mg and

(*continued*)

Table I (*continued*)

Compound (Company)	Study; Indication	Clinical Data and Status
		at this dose level produced a significant improvement in several performance measures within a cognitive screening battery (CogScreen; developed by G.G. Kay). PRX-07034 has been selected by the Treatment Units for Research on Neurocognition and Schizophrenia for a future phase II trial
SYN-114 (Synosia/ Roche)	Phase I; AD	An initial phase I trial with SYN-114 has been completed, but to date no clinical data have been reported
SUVN-502 (Suven)	Phase I; AD and obesity	Phase I studies with SUVN-502 are underway

Information on the general design and purpose of clinical trials was obtained from the U.S. National Institutes of Health http://www.clinicaltrials.gov website. Results data were obtained from press releases on the website of the sponsoring company. AD: Alzheimer's disease; PD: pharmacodynamic; PK: pharmacokinetic. (Adapted and updated from Upton *et al.*, 2008.)

GABAergic neurotransmission. It could therefore be speculated that 5-HT$_6$ receptors, localized on GABAergic interneurons, tonically modulate the activity of the GABA system. Thus, blockade of this tone effectively reduces GABA output and hence disinhibits cholinergic neurotransmission, increasing output (Woolley *et al.*, 2000, 2004). As will be discussed later the neurochemical support for this hypothesis will be reviewed, as will the evidence that this disinhibition does not exclusively modulate the cholinergic system.

Another correlation that seems to be very consistent across all small-molecule antagonist ligands is that the doses that increase cholinergic output, in *in vivo* microdialysis studies, are also roughly the doses that improve cognitive performance in normal, aged, and pharmacologically impaired rodents (see reviews by Fone, 2008; Mitchell and Neumaier, 2005; Upton *et al.*, 2008). It is therefore likely that, at least in part, 5-HT$_6$ receptor antagonist-induced efficacy in the various cognition paradigms is mediated by a modulation of cholinergic output.

B. Glutamatergic Neurotransmission

Although much of the early work did focus on the role of 5-HT$_6$ receptor antagonists in modulating cholinergic function, the first dedicated neurochemistry study demonstrated an effect on cortical glutamate output (Dawson *et al.*, 2000). These findings were further expanded upon and more fully characterized shortly afterward (Dawson *et al.*, 2001). Using *in vivo* microdialysis in the freely moving rat, it was demonstrated that SB-271046 (10 mg/kg s.c.) induced a three- and two fold increase in extracellular levels of glutamate in the dorsal lateral frontal cortex (dlFC) and hippocampus, respectively. As with other studies with SB-271046, this pharmacodynamic effect was sustained for the duration of the study (i.e. 240 min) and was maximal at 160–200 min post administration in both brain structures. Reverse microdialysis infusion of the voltage-dependent sodium channel blocker tetrodotoxin (10 μM) attenuated this SB-271046-induced effect, clearly demonstrating the impulse-dependent neuronal origin of the glutamate. As 5-HT$_6$ receptors have been suggested to enhance cholinergic function (Bentley *et al.*, 1999; Bourson *et al.*, 1995; Routledge *et al.*, 2000; Sleight *et al.*, 1999) and postsynaptic muscarinic receptor activation can induce changes in glutamate (Alcantara *et al.*, 2001; Sanz *et al.*, 1997), atropine was used in an effort to block the SB-271046-induced effects. Atropine, at doses previously shown to block 5-HT$_6$ antisense oligonucleotide-driven behaviors (3 mg/kg s.c.; Bentley *et al.*, 1999; Bourson *et al.*, 1998), administered 1 h following SB-271046 was without effect in either the dlFC or the hippocampus, suggesting that the observed increases in extracellular glutamate were not a consequence of enhanced cholinergic function. Interestingly, these enhanced glutamatergic effects were observed in the cortex and hippocampus but not within any of the other subcortical dopaminergic (i.e. striatum and nucleus accumbens) structures examined in this study, areas previously shown to be areas of relatively high 5-HT$_6$ receptor expression (Gerard *et al.*, 1996, 1997; Hamon *et al.*, 1999; Monsma *et al.*, 1993; Roberts *et al.*, 2002; Ruat *et al.*, 1993). More recently, the effect of SB-742457 on glutamatergic neurotransmission in the mPFC of rats has been shown (Chuang *et al.*, 2006). In this latest study, SB-742457 given orally produced a dose-dependent increase in extracellular levels of glutamate with a maximum effect attained at 100 min post administration (3 mg/kg p.o.). This elevated level was then maintained for the duration of the sampling regime, which again was commensurate with the PK profile of the molecule, with maximum circulating concentrations of SB-742457 being achieved at 4 h post-administration (Chuang *et al.*, 2006). Interestingly, there appeared to be a bell-shaped dose–response in this study, with SB-742457 at 10 mg/kg producing a smaller maximum increase versus that seen at 3 mg/kg. This has not been seen with previous molecules; however, SB-742457 is arguably the most potent compound tested *in vivo* to date, showing improved PK parameters and a 10-fold increase in *ex vivo* receptor occupancy ED$_{50}$'s versus SB-399885, for example (Chuang *et al.*, 2006; Hirst *et al.*, 2006). That said the exact reason for this apparent bell-shaped dose–response is not clear. Taken together these data do suggest that the increase in forebrain glutamate is a 5-HT$_6$ receptor antagonist-mediated effect as structurally diverse molecules have reproduced these findings. More recently, a number of agonist ligands have been reported in the literature. These molecules have greatly contributed to our understanding of the interplay between the 5-HT$_6$ receptor and the complex interconnecting neurochemical systems. Interestingly, the two recently reported agonists, WAY-181187 and WAY-208466, did not alter basal levels of glutamate in any brain region examined in the study reported by Schechter *et al.* (2008), even following 14 days of chronic administration of WAY-208466 (10 mg/kg s.c., q.d.). From experience, observing alterations in basal levels of glutamate sampled by *in vivo* microdialysis is notoriously difficult, particularly when that change is a decrease. The majority of glutamate sampled by microdialysis is generally not neuronal in origin and, as such, changes in neuronal output do not produce detectable alterations in basal extracellular concentrations. Stimulated efflux, on the other hand, which by nature is likely to be more neuronal, is much more easily modulated. Indeed the Schechter study (2008) clearly demonstrated that WAY-181187 attenuated but sodium azide (an ischemic/metabolic compromise challenge) and potassium stimulated glutamatergic output in an isolated hippocampal slice preparation. This provides further evidence that the 5-HT$_6$ receptor plays a central role in modulating forebrain glutamatergic neurotransmission.

Several lines of biochemical and behavioral data have provided functional support of these neurochemical observations. King *et al.* (2004) demonstrated that prior administration of the NMDA receptor antagonist MK 801 attenuated memory consolidation induced by the 5-HT$_6$ receptor antagonists SB-271046 and Ro 40-6790 (both given at 10 mg/kg i.p.) in the rat NOR paradigm. Furthermore, in the common marmoset SB-271046 (3 and 10 mg/kg p.o.) completely reversed an MK 801-induced deficit in a perceptual visual and visuospatial conditional discrimination task (Upton *et al.*, 2008). Thus, 5-HT$_6$ receptor antagonists appear to be able to reverse glutamatergic impairments or have their own efficacy prevented by NMDA receptor blockade, suggesting a direct interaction at the level of the glutamatergic synapse. Indeed biochemical studies have shown that repeated administration of SB-271046 (10 mg/kg p.o., b.i.d. for 7 days) produced changes in NMDA subunit expression (Marcos *et al.*, 2009). Furthermore, the work of Regan and colleagues (Chuang *et al.*, 2006; Foley *et al.*, 2008) has shown that a number of 5-HT$_6$ receptor antagonists are able to augment the polysialylation of a neuronal cell adhesion molecule (PSA-NCAM) in the dentate gyrus of rats. Glutamate is thought to promote the differentiation of neuronal progenitor cells, a process that involves the formation of immature neuronal populations, which express PSA-NCAM (Seki and Arai, 1999). This leads to enhanced interneuronal contact that is vital for increased neuronal plasticity processes such as long-term potentiation (Lüthl *et al.*, 1994) and hence is likely involved in the cognitive enhancing effects observed. Thus, taken together it would appear that cognitive enhancement of 5-HT$_6$ receptor antagonists may largely be mediated by the enhanced excitatory neurotransmission in forebrain structures and subsequent downstream changes in plasticity. The underlying mechanism of action behind how 5-HT$_6$ receptors modulate glutamatergic output is again somewhat paradoxical but has been speculated to be similarly to that mooted for cholinergic enhancement, that is, a disinhibition effect via modulation of GABAergic neurotransmission.

C. GABAergic Neurotransmission

A relationship between the 5-HT$_6$ receptor and the GABA system has been speculated upon for some time. Localization of 5-HT$_6$ receptors on GABAergic spiny neurons has been demonstrated (Gerard *et al.*, 1997) and colocalization with glutamic acid decarboxylase (GAD) immunoreactivity (the synthetic enzyme for GABA; Ward and Dorsa, 1996; Woolley *et al.*, 2000, 2004) also shown. This has led to the hypothesis that much of the neurochemical effects induced by antagonists have been a result of the blockade of a tonic serotonergic drive on GABAergic neurons and a subsequent disinhibition of other neuronal phenotypes. However, with the recent identification and characterization of selective 5-HT$_6$ receptor agonist ligands, this hypothesis has gained much confirmatory neurochemical support. Schechter *et al.* (2008) described two new selective small-molecule 5-HT$_6$ receptor agonists WAY-181187 and WAY-208466 (WAY-466; both molecules have been previously reported in abstract form; Beyer *et al.*, 2005; Schechter *et al.*, 2004). Administration of WAY-181187 (3–30 mg/kg s.c.) produced dose-related increases in extracellular levels of GABA in the dlFC, striatum, dorsal hippocampus, and amygdala of the rat (with no effects seen in the nucleus accumbens or thalamus). Maximum increases were observed at the 30 mg/kg dose and were two- to three fold above preinjection basal levels across all brain structures examined. This maximum increase was generally observed approximately 60 min postdose and was then maintained for the remainder of the study (i.e., 240 min postdose). This phenomenon occurred across all brain regions examined with the exception of the striatum where there was a steady increase throughout the sampling period; why this profile was observed is not clear. The maximal effect in the dlFC was fully attenuated by preadministration of SB-271046 (10 mg/kg s.c.) but the antagonist was without effects in its own right. One may have expected a decrease in GABA with the antagonist; however, this may be a technical issue, since levels of GABA sampled using overflow techniques, such as *in vivo* microdialysis, are generally quite low in structures such as the cortex (as stated in the Schechter manuscript) and hence measuring a robust decrease tends to be very difficult. This is less of an issue in structures such as the amygdala but the comparative study has not been reported. The effect of WAY-208466 was examined following a period of chronic administration (10 mg/kg s.c., q.d., 14 days), presumably to examine whether prolonged agonist exposure induced any tachyphylaxis. The agonist-induced increases in GABA in the dlFC were maintained and actually showed evidence of augmentation following 14 days of administration, suggesting that there was no desensitization of the 5-HT$_6$ receptor-induced effects. The presumed downstream effects of these changes in extracellular GABA on other neurotransmitter systems will be highlighted in the next section. However, a recent report also demonstrated that WAY-181187 was able to enhance inhibitory postsynaptic currents in the hippocampus of rats, a direct correlate of enhanced GABAergic function (West *et al.*, 2009). These observations do serve to strengthen the hypothesis that much of the neurochemical changes induced by modulating the 5-HT$_6$ receptor, and thus the range of subsequent ligand-induced changes in behavior, may indeed be a complex interplay with the GABA system playing a very central role.

D. Monoaminergic Neurotransmission

The monoaminergic systems have obviously been a focus for much neuropsychiatric research for many years. At least some of the older antidepressant type molecules (e.g., amoxapine, amitriptyline, and mianserin) and many of the atypical antipsychotics (e.g., clozapine, olanzapine, and sertindole) show moderate affinity for the 5-HT$_6$ receptor (Roth *et al.*, 1994). These types of molecules also modulate many monoaminergic systems, at least in preclinical studies (Ichikawa *et al.*, 2002a, 2002b; Kuroki *et al.*, 1999). It now seems logical that these effects may, at least in part, be mediated by this 5-HT$_6$ receptor affinity and thus a number of groups have examined this hypothesis. The effects of SB-271046 have been investigated across a number of brain structures using *in vivo* microdialysis in the rat. The initial study by Dawson *et al.* (2001) showed very little effect of SB-27016 on dopamine, noradrenaline, or 5-HT efflux across a range of brain structures, that is, dlFC, striatum, hippocampus, or nucleus accumbens. Although there was perhaps a trend to increase dopamine in the dlFC, Li *et al.* (2007) also showed a similar lack of effect of SB-399885 (3 and 10 mg/kg s.c.) on dopamine in the mPFC but a small increase in the hippocampus. In contrast, Hirst *et al.* (2003) showed that SB-399885 (1–10 mg/kg p.o.) produced a dose-related increase in mPFC dopamine and noradrenaline. No temporal profile of effect was shown but there did appear to be somewhat of a bell-shaped dose–response in this study with 3 mg/kg producing maximal effect; this was seen previously with SB-742457 (Chuang *et al.*, 2006) on glutamate changes but again no clear cause was suggested, so the reason for this remains unclear. A subsequent study (Lacroix *et al.*, 2004) examined the effect of SB-271046 in the mPFC of the rat. Here SB-271046 (1–10 mg/kg p.o.) also produced a significant increase in extracellular levels of dopamine and noradrenaline but not 5-HT. These increases were three fold above preinjection basal levels for both transmitters and reached a maximum at 120 min post administration of 10 mg/kg p.o. Interestingly, the effect on monoamines seemed to dissipate when compared to temporal profile of previous profiles seen with glutamate and ACh changes. The exact reason for this is not clear, although previous studies largely used s.c. dosing of SB-271046 versus p.o. This may have resulted in higher drug exposures; thus, the PK profile, and hence pharmacodynamic effect, may have been prolonged particularly given SB-270146's poorer oral PK profile. The agonist WAY-181187 was also examined for its effect on monoaminergic neurotransmission across six different brain structures in the Schechter study (2008). Here WAY-181187 (3–10 mg/kg s.c.) produced a decrease in extracellular levels of dopamine (at the 30 mg/kg dose) in the dlFC and striatum, and 5-HT (at 10 and 30 mg/kg s.c.) in the dlFC. All effects were attenuated by pretreatment with SB-271046 (10 mg/kg s.c.), demonstrating that these were 5-HT_6 receptor driven. Furthermore, local infusion of the $GABA_A$ receptor antagonist bicuculline (10 μM for 60 min postagonist administration) into the dlFC also fully attenuated the dopamine and 5-HT effects. These data therefore suggest that these decreases in monoamine levels were likely driven by the agonist-induced increases in extracellular levels of GABA seen with WAY-181187. Two studies both from the Dawson labs (Dawson and Li, 2003; Dawson *et al.*, 2003) looked at the combination of SB-271046 with a range of noradrenaline- and/or serotonin-based antidepressant molecules and the psychostimulant amphetamine. SB-271046 (10 mg/kg s.c.) had no effect on systemically administered desipramine-, fluoxetine-, or venlafaxine-induced changes in monoamine neurotransmission in the dlFC (Dawson and Li, 2003). One may expect that coadministration of serotonergic enhancers such as fluoxetine and venlafaxine could enhance the serotonergic tone at the 5-HT_6 receptor and thus an antagonist may have a more pronounced effect, but this was not the case, at least within the parameters measured. Interestingly, fluoxetine and venlafaxine did attenuate the SB-271046-induced increases in glutamate. The authors speculated that this may be due to enhanced serotonergic drive on the inhibitory 5-HT_{1A} heteroreceptors, which are located on glutamate neurons of the cortex. Just as an aside, blockade of this particular target has also been mooted to be a cognitive enhancing strategy (for review see Schechter *et al.*, 2002) but with limited clinical success (for review see Dawson and Bromidge, 2008). A further combination study (Li *et al.*, 2007) examined the influence of SB-399885 on the neurochemical effects of risperidone and haloperidol. SB-399885 (3 mg/kg s.c.) significantly potentiated the ability of the typical antipsychotic drug haloperidol (a D_2 receptor antagonist; 0.1 mg/kg s.c.) to increase dopamine efflux in the hippocampus but not the mPFC. The atypical antipsychotic drug risperidone (a multiaffinity monoamine receptor antagonist, which lacks 5-HT_6 receptor antagonist properties; 0.1, 0.3, and 1.0 mg/kg s.c.) produced a bell-shaped dose–response effect on dopamine efflux in the mPFC and hippocampus. SB-399885 potentiated risperidone (1.0 mg/kg s.c.)-induced dopamine efflux in both regions. These data provided additional evidence in support of the idea that the 5-HT_6 receptor antagonist activity of some atypical antipsychotics (Roth *et al.*, 1994) may be contributing to their activity.

Further studies have examined the combination of 5-HT_6 receptor antagonists with the psychostimulant amphetamine. It had been previously shown that blockade of the 5-HT_6 receptor, with SB-271046, had no effect on striatal monoaminergic neurotransmission (Dawson *et al.*, 2001), suggesting no tonic role of this receptor in regulating neurochemical output in this brain structure, despite high striatal expression (Gerard *et al.*, 1997) of the receptor. SB-271046 (10 mg/kg s.c.) augmented dopamine output induced by amphetamine (0.3 mg/kg s.c.) and also increased extracellular 5-HT levels despite amphetamine not having an effect on this neurotransmitter in its own right. This augmentation was not observed, however, when amphetamine was infused into the striatum by reverse microdialysis. This suggested that the level of interaction is remote to the striatal region sampled. The authors speculated that once again the interaction may be at the level of the striatal GABAergic spiny neurons (Gerard *et al.*, 1997) that play a role in directly and indirectly regulating nigrostriatal dopaminergic circuitry. Frantz *et al.* (2002) previously showed that an alternative antagonist SB-258510 produced an increase in amphetamine-induced locomotor behavior and augmented the amphetamine-induced increases in PFC dopamine. Furthermore, Ashby and co-workers (Minabe *et al.*, 2004) demonstrated, using *in vivo* electrophysiology, that chronic administration of SB-271046 (1–10 mg/kg p.o., o.d. for 21 days) selectively (vs. the ventral tegmental area) increased the number of spontaneously active substantia nigra dopaminergic neurons. These data further demonstrated that the 5-HT_6 receptor may play a role in regulating nigrostriatal dopaminergic output and cell firing. The reason for the 5-HT_6 receptor antagonist-induced enhancement of amphetamine on 5-HT levels is less clear and the authors suggested that the changes may be a result of other SB-271046-induced changes, for example, enhanced cortical glutamate feedback to the dorsal raphe inputs to the striatum, but of course this is speculative.

The demonstrated regulation of dopamine output is also not a phenomenon localized to the classically recognized dopamine regions or even the brain. A recent paper has shown that SB-271046 was able to elevate extracellular overflow from guinea pig cochlea, in what appeared to be a GABAergic disinhibitory manner (Doleviczeny *et al.*, 2008).

Taken as a whole it is quite clear that the 5-HT_6 receptor plays a central role in modulating multiple neurotransmitter systems across multiple brain structures and neurocircuitry. This central role is largely due to the receptor's ability to regulate the inhibitory input onto multiple systems. The localization (discussed in depth in previous chapters and re-iterated here) of the 5-HT_6 receptor on presumed GABAergic spiny neurons (Gerard *et al.*, 1997; Hamon *et al.*, 1999) and colocalization with GAD immunoreactivity (Woolley *et al.*, 2000, 2004), together with recent observations with direct agonist stimulation and enhanced GABAergic output with the selective agonist ligands, has greatly strengthened the evidence in support of this hypothesis. Figure 1 schematically speculates on how these interactions may

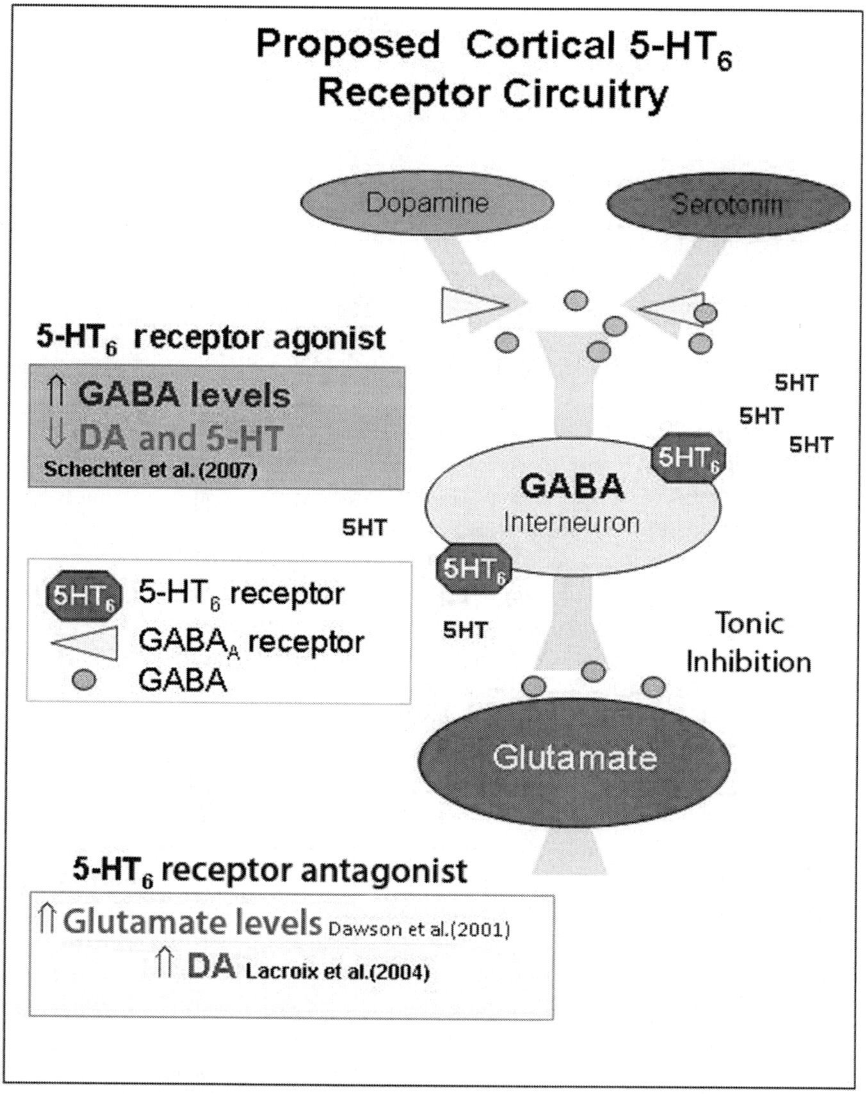

FIG. 1. The proposed neurochemical circuitry for 5-HT$_6$ receptors to influence both GABA and glutamate neurotransmission. (Figure adapted from Schechter *et al.*, 2008.) (For interpretation of the references to color in this figure legend, the reader is referred to the Web version of this chapter.)

occur in cortical and hippocampal structures. There are clearly unknown details as to how exactly the 5-HT$_6$ receptor regulates specific neurocircuitry but with the newly discovered ligands and, in particular, the receptor agonist ligands this area may yet become clearer in the future.

III. Therapeutic Implications

A. Cognitive Enhancement: Alzheimer's Disease

The role of 5-HT in cognitive processing and performance has been known for sometime (for review see Ramirez *et al.*, 2005). The 5-HT_6 receptor is one of the receptors, within the serotonergic subfamily, which has seen the most focus in this regard. From the early studies examining the effects of the antisense oligonucleotide on cognitive performance in rat water maze (Bentley *et al.*, 1997) to the extensive studies examining the selective small-molecule receptor antagonist in many diverse models of rodent and primate cognition and learning (for review see Fone, 2008; Mitchell and Neumaier, 2005; Schrieber *et al.*, 2006; Upton *et al.*, 2008; Woolley *et al.*, 2004), it has been clear that blockade of 5-HT_6 receptor-mediated neurotransmission leads to an enhanced cognitive performance in preclinical species. Interestingly, and somewhat paradoxically, it appears that 5-HT_6 receptor agonists may also be procognitive in some preclinical models of cognitive flexibility (i.e., attentional set shifting; Burnham *et al.*, 2010) and recognition memory (i.e., NOR; Kendall *et al.*, 2010). However, this was not observed in social recognition where an agonist was actually amnesic (Loiseau *et al.*, 2008), which perhaps is in more agreement with the attenuation of hippocampal LTP observed by West *et al.* (2009). As a whole from all of the data reviewed here, it would appear that the underlying mechanism of action of 5-HT_6 receptor on learning and memory processes may be via a modulatory role on multiple neurotransmitter systems.

Enhanced cholinergic output has been demonstrated with multiple chemically diverse 5-HT_6 receptor antagonists (see above). Cholinergic function has been demonstrated to play a key role in various learning and memory processes (Perry *et al.*, 1999). Indeed blockade of cholinergic function, using nonselective antagonists such as scopolamine, produces a characteristic amnesic effect across multiple species including man. However, probably the most compelling supportive data for the role of ACh in learning and memory come from the observations that normal aging and diseases such as Alzheimer's disease (AD) appear to be primarily disorders of reduced cortical cholinergic function (Terry and Buccafusco, 2003). Further, the observation that acetylcholine esterase inhibitors, which produce a global increase in ACh levels via attenuation of neurotransmitter catabolism, are currently still the frontline therapeutics for diseases that clinically manifest with symptoms of cognitive decline. Thus, a molecule that can enhance cholinergic output should have potential as a therapeutic for the treatment of diseases with an associated cognitive decline, such as AD. Indeed a number of 5-HT_6 receptor antagonists have entered the clinic for the treatment of AD (Table I) and these molecules have shown some clinical efficacy (for review see

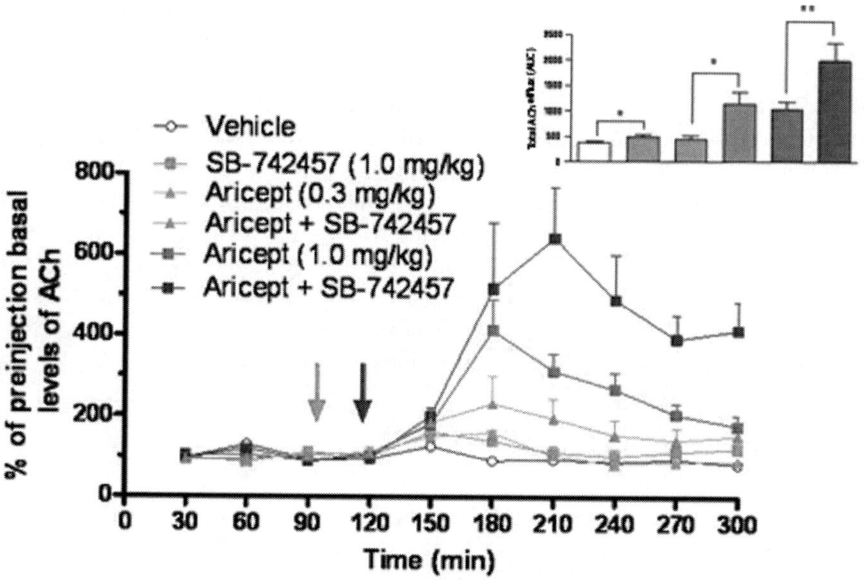

FIG. 2. Combination of SB-742457 and aricept produced an additive augmentation of extracellular levels of acetylcholine in the rat mPFC. *In vivo* microdialysis in the freely moving rat. SB-742457 (1.0 mg/kg) and aricept (either 0.3 or 1.0 mg/kg) were dosed separately via the p.o. route (SB-742457 followed by aricept). $*P < 0.05$; $**P < 0.01$. (For interpretation of the references to color in this figure legend, the reader is referred to the Web version of this chapter.)

Upton *et al.*, 2008). Interestingly based on the mechanism of action of the 5-HT$_6$ receptor, that is, removal of the tonic inhibitory drive on the cholinergic system, one may hypothesize that an antagonist may be useful as an adjunctive to acetylcholine esterase inhibitors. There are emerging preclinical data to support this hypothesis; Marcos *et al.* (2008) demonstrated that SB-271046 enhanced the efficacy of galanthamine in the Morris water maze task in rats. Furthermore, neurochemically we have demonstrated that SB-742457 can augment the aricept-induced increases in cortical ACh output as measured by microdialysis (Fig. 2). This adjunctive utility is currently being tested clinically with SB-742457 and SGS-518 (Table I), so it will be interesting to see the clinical utility of this adjunctive approach if and when data are released.

Given the complexity and diversity of the cognitive and behavioral abnormalities seen in AD, it is highly likely that these symptoms are a consequence of diverse and interactive dysfunctions of multiple neurotransmitter systems. This is supported by the observations that cholinesterase inhibitors are at best only partially efficacious in treating AD (Courtney *et al.*, 2004). Thus, the effect of 5-HT$_6$ receptor antagonists on multiple systems, including glutamate and dopamine, provides hope that these molecules may have a therapeutic impact on multiple facets of the cognitive and behavioral symptoms of AD. Both of these systems have been demonstrated to be dysfunctional at various stages in AD (Francis, 2009; Reeves *et al.*, 2010) and to play functional roles in modulating various integrated learning and memory processes (Francis, 2008; Levy and Goldman-Rakic, 2000). Certainly within the preclinical arena 5-HT_6 receptor antagonists are able to reverse not only cholinergic deficits but also glutamatergic impairments in various cognition paradigms in rodent and primate species (for reviews see Fone, 2008; Schrieber *et al.*, 2006; Upton *et al.*, 2008; Woolley *et al.*, 2004). Thus, we await further clinical data on the current (see Table I) and future clinical molecules in Alzheimer's populations to see whether efficacy is further proven and possibly enhanced versus current first-line medications.

B. Antipsychotics: Schizophrenia

Schizophrenia is a complex and highly heterogeneous disorder for which there is no common identifiable pathology and for which the etiology is poorly understood. That said the theories behind the manifestation of the various symptoms of the disease are neurochemical in nature and largely based around dopamine and glutamate dysfunction (Carlsson, 2000; Carlsson *et al.*, 2001). Conceptually, hyperactivity of the mesolimbic dopaminergic pathway is believed to cause the positive symptoms of the disease and the negative and cognitive domains are believed to arise from hypoactivity of the mesocortical dopaminergic pathways. However, it is now believed that schizophrenic symptoms reflect complex dysfunctional interplay between dopaminergic and cortical hypoglutamatergic deficits (Carlsson *et al.*, 2001; Laruelle *et al.*, 2003; Lisman *et al.*, 2008).

As mentioned previously several of the current atypical antipsychotics, for example, clozapine and olanzapine, exhibit nM affinity for the 5-HT_6 receptor (Roth *et al.*, 1994). This has led to the hypothesis that molecules targeted at this receptor may have antipsychotic efficacy. This is possibly an oversimplification, however, as atypical antipsychotics have a multifaceted pharmacology having affinity for a number of aminergic receptors. This "rich pharmacology" no doubt contributes to their efficacy and also the side effects and other issues. Preclinical neurochemistry studies have shown that molecules such as clozapine and olanzapine do have a characteristic monoaminergic signature across cortical and subcortical brain structures (Ichikawa *et al.*, 2002a, 2002b; Kuroki *et al.*, 1999; Shilliam and Dawson, 2005). The characteristic cortical (largely mPFC) neurochemical profile of the atypical antipsychotics is an increase in dopamine, noradrenaline, and ACh with no effect on 5-HT (Ichikawa *et al.*, 2002a, 2002b; Kuroki *et al.*, 1999; Shirazi-Southall *et al.*, 2002). Although a direct comparison has never been reported, the composite profile of molecules such as SB-271046 and SB-399885 would suggest that 5-HT$_6$ receptor antagonism produces a qualitative profile in the mPFC (Hirst *et al.*, 2006; Lacroix *et al.*, 2004) similar to that of an atypical antipsychotic. It also appears that adjunctive usage may be possible, since SB-399885 can augment the neurochemical effects of antipsychotics that do not possess significant 5-HT$_6$ receptor affinity (Li *et al.*, 2007). In addition to the monoaminergic effects, 5-HT$_6$ receptor antagonists enhance cortical and hippocampal glutamate levels (Dawson *et al.*, 2000, 2001), which, in our hands, atypical antipsychotics, such as clozapine, do not (unpublished observations). Since the current neurochemical theories of schizophrenia hypothesize that a dysfunction in dopamine and glutamate in the cortex may contribute to some, if not all, of the symptoms seen in schizophrenia, simplistically one may therefore conclude that 5-HT$_6$ receptor antagonists may have utility as antipsychotics. One would at least expect that the cortical enhancement in dopamine, ACh, and glutamate should improve the cognitive deficits seen in schizophrenia. In fact in a phase IIa clinical trial, SGS-518 was well tolerated and produced a dose-dependent, significant impact on the Brief Assessment of Cognition in Schizophrenia scores in a small population (20) of antipsychotic stabilized patients with schizophrenia (Lesem, 2007). It should be noted, however, that the molecule has subsequently been terminated for this indication but is still in development for AD. However, an efficacious antipsychotic would need to address the more florid positive symptoms of the disease, which probably means having a neurochemical impact on the aberrant subcortical dopaminergic systems (Carlsson, 2000; Carlsson *et al.*, 2001). Acute treatments with atypical antipsychotics produce, to greater or lesser degrees, an increase in mesolimbic and mesocortical dopamine output (Ichikawa *et al.*, 2002a, 2002b; Kuroki *et al.*, 1999; Shilliam and Dawson, 2005). However, upon chronic administration the mesolimbic output is attenuated, as shown by electrophysiological and neurochemical studies (Blackburn *et al.*, 2002; Di Giovanni *et al.*, 1998; Shilliam and Dawson, 2005; Stockton and Rasmussen, 1996). This attenuation is thought to be mesolimbic pathway specific versus nigrostriatal, for atypical antipsychotic drugs but not for typical antipsychotics such as haloperidol. This difference is thought to be the reason why antipsychotics have a delay to therapeutic onset of efficacy and why atypical antipsychotics have a reduced propensity to induce extrapyramidal side effects. The reports highlighting the acute neurochemical effects of 5-HT$_6$ receptor antagonists suggest that there is no tonic effect on either nigrostriatal or mesolimbic dopaminergic output (Dawson *et al.*, 2001). Adjunctive treatment to amphetamine has demonstrated that 5-HT$_6$ receptor antagonism can augment amphetamine-induced striatal (Dawson *et al.*, 2003) and cortical (but not nucleus accumbens; Frantz *et al.*, 2002) dopamine output. The Frantz study (2002) concurrently showed that SB-258510 also augmented amphetamine-induced locomotor behavior, which would support the enhanced neurochemical output (Dawson *et al.*, 2003). Furthermore, electrophysiology studies, examining the effects of chronic administration of SB-271046 (Minabe *et al.*, 2004), suggest that there is a selective increase in nigrostriatal activity, that is, increased dopaminergic cell firing in the substantia nigra pars compacta. Taken together these data suggest that blockade of the 5-HT_6 receptor would have limited utility in the treatment of the positive symptoms of schizophrenia. Selective augmentation of nigrostriatal pathways *per se* and an augmentory effect on hyperactive nigrostriatal output (as in the case with amphetamine) may have an impact on positive symptoms and/or EPS liability in schizophrenic patients. This is of course speculation and in fact a role of this receptor in dependency has been speculated upon (Ferguson *et al.*, 2008), which may actually suggest the contrary. Although some clinical evaluation in schizophrenics has been performed, i.e. the SGS-518 study that examined cognitive effects (Table I), it is impossible to draw any clear conclusions in this regard. This study was performed in stabilized patients, that is, as an add-on to D_2 blocking agents, an activity that would ameliorate any clear behavioral outcome of elevations in dopamine. Similarly all antipsychotics that possess 5-HT_6 receptor affinity are D_2 receptor antagonists; so again any potentially detrimental impact of 5-HT_6 receptor-induced changes in subcortical dopamine system would be masked. There are no reported clinical studies that have demonstrated antipsychotic efficacy of 5-HT_6 receptor antagonists as stand-alone treatments. The preclinical evidence, in models thought to be predictive of antipsychotic efficacy, are also limited and those studies that have examined this are very mixed in their conclusions (Leng *et al.*, 2003; Mitchell and Neumaier, 2008; Pouzet *et al.*, 2002; Schreiber *et al.*, 2007). It would therefore seem from available data that 5-HT_6 receptor antagonism may have some utility in the treatment of cognitive deficits in schizophrenia but the supportive evidence for utility against the other symptom domains of the disease is somewhat limited.

C. Anxiety/Depression

The introduction of novel 5-HT_6 receptor agonists (Cole *et al.*, 2007) has provided strong supporting evidence to the immunohistochemical studies (Gerard *et al.*, 1997; Hamon *et al.*, 1999; Woolley *et al.*, 2001, 2004), suggesting that 5-HT_6 receptors play a central role in modulating GABAergic neurotransmission across the brain. In this regard, neurochemical studies have demonstrated that both WAY-181187 and WAY-208466 consistently elevate levels of GABA in multiple brain regions associated with anxiety, including frontal cortex, hippocampus, and amygdala (Schechter *et al.*, 2008). Current therapeutic agents for the treatment of anxiety disorders include benzodiazepines and serotonin-specific reuptake inhibitors (SSRIs), both of which act either directly or indirectly to modulate GABAergic neurotransmission. Benzodiazepines act as positive allosteric modulators of the GABA$_A$ receptor/Cl$^-$ ion channel complex, and thus enhance GABAergic signaling. SSRIs also may enhance levels of GABA as predicted from recent imaging studies in humans (Sanacora *et al.*, 2002). Additionally, both WAY-181187 and WAY-208466 attenuate stimulated glutamate efflux in brain slices (Schechter *et al.*, 2008). Under stressful situations, glutamatergic neurotransmission may increase in cortical and limbic systems and may be associated with anxiety symptoms as well as hippocampal atrophy. These glutamate attenuating effects of 5-HT$_6$ receptor agonists may also be beneficial in obsessive–compulsive disorder, which may also involve increased levels of glutamate and dopamine (Carlsson *et al.*, 2001; Stein, 2000). Electrophysiological studies have shown that chronic administration of a 5-HT$_6$ receptor agonist decreased the basal firing rate of the substantia nigra dopaminergic cells (Schechter *et al.*, 2005), which correlates well with the previous data with SB-271046 (Minabe *et al.*, 2004) and suggests that agonists may also attenuate aberrant dopamine output. Supporting behavioral observations have shown that 5-HT$_6$ receptor agonists are effective acutely in preclinical models of anxiety/depression (Carr *et al.*, 2010; Svenningsson *et al.*, 2007) and schedule-induced polydipsia model, indicative of a rapid-onset anti-OCD-like effect (Schechter *et al.*, 2008). Furthermore, the 5-HT$_6$ receptor agonist LY586713 upregulates BDNF mRNA in the hippocampus following either acute or short-term (4 days) treatment effects that can be blocked by a 5-HT$_6$ receptor antagonist (De Foubert *et al.*, 2007). Antidepressants, including SSRIs, have been shown to upregulate BDNF gene expression (Russo-Neustadt and Chen, 2005). However, in contrast to the observations with 5-HT$_6$ receptor agonists, the Wesolowska group has shown that SB-399885 consistently produces anxiolytic/antidepressant efficacy in preclinical models (Wesolowska, 2007, 2008; Wesolowska and Nikiforuk, 2007). Clearly further work needs to be done to more fully evaluate 5-HT$_6$ receptor modulators in anxiety/depression/OCD type disorders. 5-HT$_6$ receptor agonists seem to have the most supportive evidence to date at least as anxiolytics but these ligands are still early in the drug development process, so we await emerging data with interest.

IV. Conclusions

Since the discovery of the 5-HT$_6$ receptor and the development of small-molecule antagonist and agonist ligands, neurochemical and localization studies have clearly demonstrated a central role for this receptor in modulating GABAergic neurotransmission across a wide range of brain structures and networks. As a consequence of this central role, 5-HT$_6$ receptor ligands can subsequently produce a diverse series of changes in monoaminergic, cholinergic, and excitatory amino acid neurotransmission (as discussed herein). These neurochemical data have provided supportive data to suggest a number of therapeutic avenues that may be pursued for antagonist and more recently agonist ligands. Preclinical behavioral studies have concurrently provided supportive evidence for targeting relevant CNS disorders. Improvements in cognitive performance, learning, and memory have largely been the focus of this area thus far and clinical evaluation has reached phase II for the furthest developed molecules. However, as data continue to emerge other areas may also be pursued (some reviewed here and in later chapters, e.g., obesity) in the future, particularly with the emergence of the newer 5-HT$_6$ receptor agonist ligands. This receptor still offers much potential as a therapeutic intervention point for a range of CNS-related disorders and we await the outcomes of current and future clinical evaluations with interest and hope.

References

Alcantara, A.A., Mrzljak, L., Jakab, R.L., Levey, A.I., Hersch, S.M., and Goldman-Rakic, P.S. (2001). Muscarinic m1 and m2 receptor proteins in local circuit and projection neurons of the primate striatum: anatomical evidence for cholinergic modulation of glutamatergic prefronto-striatal pathways. *J. Comp. Neurol.* **434**, 445–460.

Bentley, J.C., Bourson, A., Boess, F.G., Fone, K.C.F., Marsden, C.A., Petit, N., and Sleight, A.J. (1999). Investigation of stretching behaviour induced by the selective 5-HT$_6$ receptor antagonist, Ro04-6790, in rats. *Br. J. Pharmacol.* **126**, 1537–1542.

Bentley, J.C., Sleight, A.J., Marsden, C.A., and Fone, K.C.F. (1997). 5-HT$_6$ antisense oligonucleotide i.c.v. affects rat performance in the water maze and feeding. *J. Psychopharmacol.* **11**, A64.

Beyer, C.E., Smith, D.L., Zhang, G., Li, P., Lin, J.R., Stock, J.W., Ellinboe, J.W., Bernotas, R., Cole, D. C., Dawson, L.A., and Schechter, L.E. (2005). WAY-181187: neurochemical profile of a novel and selective 5-HT$_6$ receptor agonist. *Eur. Neuropsychopharmacol.* **15**, S382.

Blackburn, T.P., Minabe, Y., Middlemiss, D.N., Shirayama, Y., Hashimoto, K., and Ashby Jr., C.R. (2002). Effect of acute and chronic administration of the selective 5-HT2C receptor antagonist SB-243213 on midbrain dopamine neurons in the rat: an *in vivo* extracellular single cell study. *Synapse* **46**, 129–139.

Bourson, A., Boess, F.G., Bos, M., and Sleight, A.J. (1998). Involvement of 5-HT$_6$ receptors in nigrostriatal function in rodents. *Br. J. Pharmacol.* **125**, 1562–1566.

Bourson, A., Borroni, E., Austin, R.H., Monsma, F.J., and Sleight, A.J. (1995). Determination of the role of the 5-HT$_6$ receptor in the rat brain: a study using antisense oligonucleotides. *J. Pharmacol. Exp. Ther.* **274**, 173–180.

Bromidge, S.M., Brown, A.M., Clarke, S.E., Dodgson, K., Gagewr, T., Grassam, H.L., Jeffrey, P.M., Joiner, G.F., King, F.D., Middlemiss, D.N., Moss, S.F., Newman, H., Riley, G., Routledge, C., and Wyman, P. (1999). 5-Choro-*N*-(4-methoxy-3-piperazin-1-yl-phenyl)-3-methyl-2-benzothiophenesulfonamide (SB-271046): a potent selective, and orally bioavailable 5-HT$_6$ receptor antagonist. *J. Med. Chem.* **42**, 202–205.

Burnham, K.E., Baxter, M.G., Bainton, J.R., Southam, E., Dawson, L.A., Sharp, T., and Bannerman, D.M. (2010). Activation of 5-HT$_6$ receptors facilitates attentional set-shifting. *Psychopharmacology* **208**, 13–21.

Carlsson, M.L. (2000). On the role of cortical glutamate in obsessive-compulsive disorder and attention-deficit hyperactivity disorder, two phenomenologically antithetical conditions. *Acta Psychiatr. Scand.* **102**, 401–413.

Carlsson, A., Waters, N., Holm-Waters, S., Tedroff, J., Nilsson, M., and Carlsson, M. (2001). Interactions between monoamines, glutamate and GABA in schizophrenia: new evidence. *Annu. Rev. Pharmacol. Toxicol.* **41**, 237–260.

Carr, G.V., Schechter, L.E., and Lucki, I. (2010). Antidepressant and anxiolytic effects of selective 5-HT$_6$ receptor agonists in rats. *Psychopharmacology* (Epub ahead of print) DOI: 10.1007/s00213-010-1798-7

Chuang, A.T.T., Foley, A., Pugh, P.L., Sunter, D., Tong, X., Regan, C., and Dawson, L.A *et al.* (2006). 5-HT$_6$ receptor antagonist SB-742457 as a novel cognitive enhancing agent for Alzheimer's disease. *Alzheimer's Dementia* **2**, S631–S632.

Cole, D.C., Stock, J.R., Lennox, W.J., Bernotas, R.C., Ellingboe, J.W., Boikess, S., Coupet, J., Smith, D. L., Leung, L., Zhang, G.M., Feng, X., Kelly, M.F., Galante, R., Huang, P., Dawson, L.A., Marquis, K., Rosenweig-Lipson, S., Beyer, C.E., and Schechter, L.E. (2007). Discovery of *N*1-(6-chloroimidazo[2,1-*b*][1,3]thiazole-5-sulfoyl)tryptamine as a potent, selective, and orally active 5-HT$_6$ receptor agonist. *J. Med. Chem.* **50**, 5535–5538.

Courtney, C., Farrell, D., and Gray, RAD2000 Collaborative Group(2004). Long-term donepezil treatment in 565 patients with Alzheimer's disease (AD2000): randomised double-blind trial. *Lancet* **363**, 2105–2115.

Dawson, L.A., and Bromidge, S.M. (2008). 5-HT$_1$ Receptor augmentation strategies as enhanced efficacy therapeutics for psychiatric disorders. *Curr. Top. Med. Chem.* **8**, 1008–1023.

Dawson, L.A., and Li, P. (2003). Effects of 5-HT$_6$ receptor blockade in the neurochemical outcome of antidepressant treatment in the frontal cortex of the rat. *J. Neural. Transm.* **110**, 577–590.

Dawson, L.A., Nguyen, H.Q., and Li, P. (2000). *In vivo* effects of the 5-HT$_6$ antagonist SB-271046 on striatal and frontal cortex extracellular concentrations of noradrenaline, dopamine, 5-HT, glutamate and aspartate. *Br. J. Pharmacol.* **130**, 23–26.

Dawson, L.A., Nguyen, H.Q., and Li, P. (2001). The 5-HT$_6$ receptor antagonist SB-271046 selectively enhances excitatory neurotransmission in the rat frontal cortex and hippocampus. *Neuropsychopharmacology* **25**, 662–668.

Dawson, L.A., Nguyen, H.Q., and Li, P. (2003). Potentiation of amphetamine-induced changes in dopamine and 5-HT by a 5-HT$_6$ receptor antagonist. *Brain Res. Bull.* **59**, 513–521.

De Foubert, G., O'Neill, M.J., and Zetterstrom, T.S. (2007). Acute onset by 5-HT$_6$ receptor-activation on rat brain-derived neurotrophic factor and activity-regulated cytoskeletal-associated protein mRNA expression. *Neuroscience* **147**, 778–785.

Di Giovanni, G., Di Mascio, M., Di Matteo, V., and Esposito, E. (1998). Effects of acute and repeated administration of amisulpiride, a dopamine D_2/D_3 receptor antagonist, on the electrical activity of midbrain dopaminergic neurones. *J. Pharmacol. Exp. Ther.* **287**, 51–57.

Doleviczeny, Z., Vizi, E.S., Gacsalyi, I., Pallagi, K., Volk, B., Harsing Jr., L.G., Halmois, G., Lendvai, B., and Zelles, T. (2008). 5-HT$_{6/7}$ receptor antagonists facilitate dopamine release in the cochlea via GABAergic disinhibitory mechanism. *Neurochem. Res.* **33**, 2364–2372.

Ferguson, S.M., Mitchell, E.S., and Neumaier, J.F. (2008). Increased expression of 5-HT$_6$ receptors in the nucleus accumbens blocks the rewarding but not psychomotor activating properties of cocaine. *Biol. Psychiatry* **63**, 207–213.

Foley, A.G., Hirst, W.D., Gallagher, H.C., Barry, C., Hagan, J.J., Upton, N., Walsh, F.S., Hunter, A.J., and Regan, C.M. (2008). The selective 5-HT$_6$ antagonists SB-271046 and SB-399885 potentiate NCAM PSA immunolabeling of dentate granule cells, but not neurogenesis, in the hippocampal formation of mature Wistar rats. *Neuropharmacology* **54**, 1166–1174.

Foley, A.G., Murphy, K.J., Hirst, W.D., Gallagher, H.C., Hagan, J.J., Upton, N., Walsh, F.S., and Regan, C.M. (2004). The 5-HT$_6$ receptor antagonist SB-271046 reverses scopolamine-induced consolidation of a passive avoidance task and ameliorates spatial task deficits in aged rats. *Neuropsychopharmacology* **29**, 93–100.

Fone, K.C.F. (2008). An update on the role of the 5-hydroxytryptamine$_6$ receptor in cognitive function. *Neuropharmacology* **55**, 1015–1022.

Francis, P.T. (2008). Glutamatergic approaches to the treatment of cognitive and behavioural symptoms of Alzheimer's disease. *Neurodegener. Dis.* **5**, 241–243.

Francis, P.T. (2009). Altered glutamate neurotransmission and behaviour in dementia: evidence from studies of memantine. *Curr. Mol. Pharmacol.* **2**, 77–82.

Frantz, K.J., Hansson, K.J., Stouffer, D.G., and Parson, L.H. (2002). 5-HT$_6$ receptor antagonism potentiates the behavioural and neurochemical effects of amphetamine but not cocaine. *Neuropharmacology* **42**, 170–180.

Gerard, C., El Mestkawy, S., Lebrand, C., Adrien, J., Ruat, M., Traiffort, E., Hamon, M., and Martres, M.P. (1996). Quantitative RT-PCR distribution of serotonin 5-HT6 receptor mRNA in the central nervous system of control or 5,7-dihydroxytryptramine-trated rats. *Synapse* **23**, 164–173.

Gerard, C., Martres, M.P., Lefevre, K., Miquel, M.C., Vergé, D., Lanfumey, L., Doucet, E., Hamon, M., and El Mestkawy, S. (1997). Immuno-localization of serotonin in 5-HT6 receptor-like material in the rat central nervous system. *Brain Res.* **746**, 207–219.

Hamon, M., Doucet, E., Lefevre, K., Miquel, M.C., Lanfumey, L., Insausti, R., Frechilla, D., Del Rio, J., and Vergé, D. (1999). Antibodies and antisense oligonucleotide for probing the distribution and putative functions of central 5-HT$_6$ receptors. *Neuropsychopharmacology* **21**, 68S–76S.

Hatcher, P.D., Brown, V.J., Tait, D.S., Bate, S., Overend, P., Hagan, J.J., and Jones, D.N.C. (2005). 5-HT$_6$ receptor antagonists improve performance in an attentional set shifting task in rats. *Psychopharmacology* **181**, 253–259.

Hirst, W.D., Moss, S.F., Bromidge, S.M., Riley, G., Stean, T.O., Rogers, D.C., and Sunter, D *et al.* (2003). Characterisation of SB-399885, a potent and selective 5-HT$_6$ receptor antagonist. *Soc. Neurosci. Abstr.* **29**, 576–577.

Hirst, W.D., Stean, T.O., Rogers, D.C., Sunter, D., Pugh, P., Moss, S.F., and Bromidge, S.M *et al.* (2006). SB-399885 is a potent, selective 5-HT$_6$ receptor antagonist with cognitive enhancing properties in aged rat water maze and novel recognition models. *Eur. J. Pharmacol.* **553**, 109–119.

Ichikawa, J., Dai, J., O'Laughlin, I.A., Fowler, W.L., and Meltzer, H.Y. (2002a). Atypical, but not typical, antipsychotic drugs increase cortical acetylcholine release without an effect in the nucleus accumbens or striatum. *Neuropsychopharmacology* **26**, 325–339.

Ichikawa, J., Li, Z., Dai, J., and Meltzer, H.Y. (2002b). Atypical antipsychotic drugs, quetiapine, iloperidone, and melperone, preferentially increase dopamine and acetylcholine release in rat medial prefrontal cortex: role of 5-HT$_{1A}$ receptor agonism. *Brain Res.* **956**, 349–357.

Kendall, I., Slotten, H.A., Codony, X., Burgueno, J., Pauwels, P.J., Vela, J.M., and Fone, K.C. (2010). E-6801, a 5-HT$_6$ receptor agonist, improves recognition memory by combined modulation of cholinergic and glutamatergic neurotransmission in the rat. *Psychopharmacology* (Epub ahead of print) DOI: 10.1007/s00213-010-1854-3

King, M.V., Sleight, A.J., Woolley, M.L., Topham, I.A., Marsden, C.A., and Fone, K.C.F. (2004). 5-HT$_6$ receptor antagonists reverse delay-dependent deficits in novel discrimination by enhancing consolidation: an effect sensitive to NMDA receptor antagonism. *Neuropharmacology* **47**, 195–204.

Kohen, R., Metcalf, M.A., Khan, N., Druck, T., Huebner, K., Lachowicz, J.E., Meltzer, H.Y., Sibley, D.R., Roth, B.L., and Hamblin, M.W. (1996). Cloning, characterization, and chromosomal localization of a human 5-HT$_6$ serotonin receptor. *J. Neurochem.* **66**, 47–56.

Kuroki, T., Meltzer, H.Y., and Ichikawa, J. (1999). Effects of antipsychotic drugs on extracellular dopamine levels in rat medial prefrontal cortex and nucleus accumbens. *J. Pharmacol. Exp. Ther.* **288**, 774–781.

Lacroix, L.P., Dawson, L.A., Hagan, J.J., and Heidbreder, C.A. (2004). 5-HT$_6$ receptor antagonist SB-271046 enhances extracellular levels of monoamines in the rat medial prefrontal cortex. *Synapse* **51**, 158–164.

Laruelle, M., Kegeles, L., and Abi-Dargham, A. (2003). Glutamate, dopamine and schizophrenia: from pathophysiology to treatment. *Ann. N.Y. Acad. Sci.* **1003**, 138–158.

Leng, A., Ouagazzal, A., Feldon, J., and Higgins, G.A. (2003). Effect of the 5-HT$_6$ receptor antagonists Ro 04-6790 and Ro 65-7199 on latent inhibition and prepulse inhibition in the rat: comparison to clozapine. *Pharmacol. Biochem. Behav.* **75**, 281–288.

Lesem, M. (2007). A randomized, placebo-controlled phase IIa trial of SGS518 for treating cognitive impairment associated with schizophrenia. Proceedings from 11th International Congress on Schizophrenia Research.

Levy, R., and Goldman-Rakic, P.S. (2000). Segregation of working memory functions within the dorsolateral prefrontal cortex. *Exp. Brain Res.* **133**, 23–32.

Li, Z., Huang, M., Prus, A.J., Dai, J., and Meltzer, H.Y. (2007). 5-HT$_6$ receptor antagonist SB-399885 potentiates haloperidol and risperidone-induced dopamine efflux in the medial prefrontal cortex or hippocampus. *Brain Res.* **1134**, 70–78.

Liebnen, C.K.J., Blokland, A., Sik, A., Sung, E., van Nieuwenhuizen, P., and Schreober, R. (2005). The selective 5-HT$_6$ receptor antagonists Ro4368554 restores memory performance in cholinergic and serotonergic models of memory deficiency in the rat. *Neuropsychopharmacology* **30**, 2169–2179.

Lisman, J.E., Coyle, J.T., Green, R.W., Javitt, D.C., Benes, F.M., Heckers, S., and Grace, A.A. (2008). Circuit-based framework for understanding neurotransmitter and risk gene interactions in schizophrenia. *Trends Neurosci.* **31**, 234–242.

Loiseau, F., Dekeyne, A., and Millan, M.J. (2008). Pro-cognitive effects of 5-HT$_6$ receptor antagonists in the social recognition procedure in rats: implication of the frontal cortex. *Psychopharmacology* **196**, 93–104.

Lüthl, A., Laurent, J.P., Figurov, A., Muller, D., and Schachner, M. (1994). Hippocampal long-term potentiation and neural cell adhesion molecules L1 and NCAM. *Nature* **372**, 777–779.

Marcos, B., Aisa, B., Chuang, T.T., Gil-Bea, F.J., and Ramírez, M.J. (2009). Effects of chronic blockade of 5-HT$_6$ receptors on NMDA receptor subunits expression. *Synapse* **63**, 814–816.

Marcos, B., Chuang, T.T., and Ramirez, M.J. (2008). Effects of 5-HT$_6$ receptor antagonism and cholinesterase inhibition in models of cognitive impairment in the rat. *Br. J. Pharmacol.* **155**, 434–440.

Marcos, B., Gil-Bea, F.J., Hirst, W.D., Garcia-Alloza, M., and Ramirez, M.J. (2006). Lack of localization of 5-HT$_6$ receptors on cholinergic neurons: implication of multiple neurotransmitter systems on 5-HT$_6$ receptor-mediated acetylcholine release. *Eur. J. Neurosci.* **24**, 1299–1306.

Minabe, Y., Shirayama, Y., Hashimoto, K., Routledge, C., Hagan, J.J., and Ashny Jr., C.R. (2004). Effect of acute and chronic administration of the selective 5-HT$_6$ receptor antagonist SB-271046 on the activity of midbrain dopamine neurons in rats: an *in vivo* electrophysiological study. *Synapse* **52**, 20–28.

Mitchell, E.S., and Neumaier, J.F. (2005). 5-HT$_6$ receptors: a novel target for cognitive enhancement. *Pharmacol. Ther.* **108**, 320–333.

Mitchell, E.S., and Neumaier, J.F. (2008). 5-HT$_6$ receptor antagonist of emotional learning and prepulse inhibition deficits induced by apomorphine or scopolamine. *Pharmacol. Biochem. Behav.* **88**, 291–298.

Monsma Jr., F.J., Shen, Y., Ward, R.P., Hamblin, M.W., and Sibley, D.R. (1993). Cloning and expression of a novel serotonin receptor with affinity for tricyclic psychotropic drugs. *Mol. Pharmacol.* **43**, 320–327.

Perry, E., Walker, M., Grace, J., and Perry, R. (1999). Acetylcholine in mind: a neurotransmitter correlate of consciousness? *Trends Neurosci.* **22**, 273–280.

Pouzet, B., Didriksen, M., and Arnt, J. (2002). Effects of the 5-HT$_6$ receptor antagonist, SB-271046, in animal models for schizophrenia. *Pharmacol. Biochem. Behav.* **71**, 635–643.

Ramirez, M.J., Aisa, B., Garcia-Alloza, M., Gil-Bea, F.J., and Marcos, B. (2005). Involvement of the serotonergic system in cognitive and behavioral symptoms of Alzheimer's disease. *Curr. Psychiatry Rev.* **1**, 337–343.

Reeves, S., Mehta, M., Howard, R., Grasby, P., and Brown, R. (2010). The dopaminergic basis of cognitive and motor performance in Alzheimer's disease. *Neurobiol. Dis.* **37**, 477–482.

Riemer, C., Borroni, E., Levet-Trafit, B., Martin, J.R., Poli, S., Porter, R.H.P., and Bos, M. (2003). Influence of the 5-HT$_6$ receptor on acetylcholine release in the cortex: pharmacological characterization of 4-(2-bromo-6-pyrrolidin-1-ylpyridine-4-sulfonyl) phenylamine, a potent and selective 5-HT$_6$ receptor antagonist. *J. Med. Chem.* **46**, 1273–1276.

Roberts, J.C., Reavill, C., East, S.Z., Harrison, P.J., Patel, S., Routledge, C., and Leslie, R.A. (2002). The distribution of 5-HT$_6$ receptors in rat brain: an autoradiographic binding study using the radiolabelled 5-HT$_6$ receptor antagonist [^{125}I]SB-258585. *Brain Res.* **934**, 49–57.

Rogers, D.C., and Hagan, J.J. (2001). 5-HT$_6$ receptor antagonists enhance retention of a water maze task in the rat. *Psychopharmacology* **158**, 114–119.

Rogers, D.C., Robinson, T.L., Quilter, C.A., Hunter, A.I., Routledge, C., and Hagan, J.I. (1999). Cognitive enhancement effects of the selective 5-HT$_6$ antagonist SB-271046. *Br. J. Pharmacol.* **127**, 22P.

Roth, B.L., Craigo, S.C., Choudhary, M.S., Uluier, A., Monsma Jr., F.J., Shen, Y., Meltzer, H.Y., and Sibley, D.R. (1994). Binding to typical and atypical antipsychotic agents to 5-hydroxytryptamine-6 and 5-hydroxytryptamine-7 receptors. *J. Pharmacol. Exp. Ther.* **268**, 1403–1410.

Routledge, C., Bromidge, S.M., Moss, S.F., Price, G.W., Hirst, W., Newmann, H., Riley, G., Gager, T., Stean, T., Upton, U., Clarke, S.E., Brown, A.M., and Middlemiss, D.N. (2000). Characterization of SB-271046: a potent, selective and orally active 5-HT$_6$ receptor antagonist. *Br. J. Pharmacol.* **130**, 1606–1612.

Ruat, M., Traiffort, E., Arrang, J.-M., Tardivel-Lacombe, J., Diaz, J., Leurs, R., and Schwartz, J.-C. (1993). A novel rat serotonin (5-HT$_6$) receptor: molecular cloning, localization and stimulation of cAMP. *Biochem. Biophys. Res. Commun.* **193**, 268–276.

Russo-Neustadt, A.A., and Chen, M.J. (2005). Brain-derived neurotrophic factor and antidepressant activity. *Curr. Pharm. Des.* **11**, 1495–1510.

Sanacora, G., Mason, G.F., Rothman, D.L., and Krystal, J.H. (2002). Increased occipital cortex GABA concentrations in depressed patients after therapy with selective serotonin reuptake inhibitors. *Am. J. Psychiatry* **159**, 663–665.

Sanz, B., Exposito, I., and Mora, F. (1997). M1 acetylcholine receptor stimulation increases the extracellular concentrations of glutamate and GABA in the medial prefrontal cortex of the rat. *Neurochem. Res.* **22**, 281–286.

Schechter, L.E., Dawson, L.A., and Harder, J.A. (2002). The potential utility of 5-HT$_{1A}$ receptor antagonists in the treatment of cognitive dysfunction associated with Alzheimer's disease. *Curr. Pharm. Des.* **8**, 139–145.

Schechter, L.E., Lin, Q., Smith, D.L., Zhang, G., San, Q., Platt, B., Brandt, M.R., Dawson, L.A., Cole, D., Bernotas, R., Ribichaud, A., Rosenweig-Lipson, S., and Beyer, C.E. (2008). Neuropharmacological profile of novel and selective 5-HT$_6$ receptor agonists: WAY-181187 and WAY-208466. *Neuropsychopharmacology* **33**, 1323–1335.

Schechter, L.E., Ring, R.H., Beyer, C.E., Hughes, Z.A., Khawaja, X., Malberg, J.E., and Rosenzweig-Lipson, S. (2005). Innovative approaches for the development of antidepressant drugs: current and future strategies. *NeuroRx* **2**, 590–611.

Schechter, L.E., Smith, D.L., Li, P., Lin, Q., Rosenweig-Lipson, S., and Ribichaud, A *et al.* (2004). WAY-466: pharmacological profile of novel and selective 5-HT$_6$ receptor agonists. *Soc. Neurosci. Abstr.* **394**, 11.

Schrieber, R., Sleight, A., and Woolley, M.L. (2006). 5-HT$_6$ receptors as targets for the treatment of cognitive deficits in schizophrenia. In: Roth, B.L. (Ed.), *The Serotonin Receptors: From Molecular Pharmacology to Human Therapeutics*. Humana Press, New Jersey.

Schreiber, R., Vivian, J., Hedley, L., Szcepanski, K., Secchi, R.L., Zuzow, M., van Laarhoven, S., Moreau, J.-L., Martin, J.R., Sik, A., and Blokland, A. (2007). Effects of the novel 5-HT$_6$ receptor antagonist Ro4368554 in rat models for cognition and sensorimotor gating. *Eur. Neuropsychopharmacol.* **17**, 277–288.

Seki, T., and Arai, Y. (1999). Different polysialic acid-neural cell adhesion molecule expression patterns in distinct types of mossy fiber boutons in the adult hippocampus. *J. Comp. Neurol.* **410**, 115–125.

Shilliam, C.S., and Dawson, L.A. (2005). The effect of clozapine on extracellular dopamine levels in the shell subregion of the rat nucleus accumbens is reversed following chronic administration: comparison with a selective 5-HT$_{2C}$ receptor antagonist. *Neuropsychopharmacology* **30**, 372–380.

Shirazi-Southall, S., Rodriguez, D.E., and Nomikos, G.G. (2002). Effects of typical and atypical antipsychotics and receptor selective compounds on acetylcholine efflux in the hippocampus of the rat. *Neuropsychopharmacology* **26**, 583–594.

Sleight, A.J., Boess, F.G., Bos, M., Levet-Trafit, B., Riemer, C., and Bourson, A. (1998). Characterization of Ro 04-6790 and Ro 63-0563: potent and selective antagonists at human and rat 5-HT$_6$ receptors. *Br. J. Pharmacol.* **1245**, 556–562.

Sleight, A.J., Consolo, S., Martin, J.R., Boes, M., Boess, F.G., Bentley, J.C., and Bourson, A. (1999). 5-HT$_6$ receptors: functional correlates and potential therapeutic indications. *Behav. Pharmacol.* **10** (Suppl. 1); S86–S87.

Stein, D.J. (2000). Neurobiology of the obsessive-compulsive spectrum disorders. *Biol. Psychiatry* **47**, 296–304.

Stockton, M.E., and Rasmussen, K. (1996). Electrophysiological effects of olanzapine, a novel atypical antipsychotic, on A9 and A10 dopamine neurons. *Neuropsychopharmacology* **14**, 97–104.

Svenningsson, P., Tzavara, E.T., Qi, H., Carruthers, R., Witkin, J.M., Nomikons, G.G., and Greengard, P. (2007). Biochemical and behavioural evidence for antidepressant-like effects of 5-HT$_6$ receptor stimulation. *J. Neurosci.* **27**, 4201–4209.

Terry Jr., A.V., and Buccafusco, J.J. (2003). The cholinergic hypothesis of age and Alzheimer's disease-related cognitive deficits: recent challenges and their implications for novel drug development. *J. Pharmacol. Exp. Ther.* **306**, 821–827.

Upton, N., Chuang, T.T., Hunter, A.J., and Virley, D.J. (2008). 5-HT$_6$ receptor antagonists as novel cognitive enhancing agents for Alzheimer's disease. *Neurotherapeutics* **5**, 458–469.

Ward, R., and Dorsa, D.M. (1996). Colocalization of serotonin receptor subtypes 5-HT$_{2A}$, 5-HT$_{2C}$, and 5-HT$_6$ with neuropeptides in rat striatum. *J. Comp. Neurol.* **370**, 405–414.

Ward, R.P., Hamblin, M.W., Lachowicz, J.E., Hoffman, B.J., Sibley, D.R., and Dorsa, D.M. (1995). Localization of serotonin subtype 6 receptor messenger RNA in the rat brain by *in situ* hybridization histochemistry. *Neuroscience* **64**, 1105–1111.

Wesolowska, A. (2007). Study into a possible mechanism responsible for the antidepressant-like activity of the selective 5-HT$_6$ receptor antagonist SB-399885 in rats. *Pharmacol. Rep.* **59**, 664–671.

Wesolowska, A. (2008). The anxiolytic-like effect of the selective 5-HT$_6$ receptor antagonist SB-399885: the impact of benzodiazepines receptors. *Eur. J. Pharmacol.* **580**, 355–360.

Wesolowska, A., and Nikiforuk, A. (2007). Effects of the brain-penetrant and selective 5-HT$_6$ receptor antagonists SB-3999885 in animal models of anxiety and depression. *Neuropharmacology* **52**, 1274–1283.

West, P.J., Marcy, V.R., Marino, M.J., and Schaffhauser, H. (2009). Activation of the 5-HT$_6$ receptor attenuates long-term potentiation and facilitates GABAergic neurotransmission in rat hippocampus. *Neuroscience* **164**, 692–701.

Woolley, M.L., Bentley, J.C., Sleight, A.J., Marsden, C.A., and Fone, K.C.F. (2001). A role for $5\text{-}HT_6$ receptors in retention of spatial learning in the Morris water maze. *Neuropharmacology* **41**, 210–219.

Woolley, M.L., Marsden, C.A., and Fone, K.C.F. (2004). $5\text{-}HT_6$ receptors. *Curr. Drug Targets* **3**, 59–89.

Woolley, M. L., Marsden, C. A., Sleight, A. J., and Fone, K. C. F. (2000). Co-localisation of $5\text{-}HT_6$ receptors on GABAergic but not cholinergic neurones in the adult rat brain. *Brain Res. Bull*: Serotonin: from the molecule to the clinic, p. 95.

Zhang, M.Y., Hughes, Z.A., Kerns, E.H., Lin, Q., and Beyer, C.E. (2007). Development of a liquid chromatography/tandem mass spectrometry method for the quantitation of acetylcholine and related neurotransmitters in brain microdialysis samples. *J. Pharm. Biomed. Anal.* **44**, 586–593.

5-HT$_6$ RECEPTOR MEMORY AND AMNESIA: BEHAVIORAL PHARMACOLOGY – LEARNING AND MEMORY PROCESSES

Alfredo Meneses[1], Georgina Pérez-García[1], Teresa Ponce-Lopez[1] and Carlos Castillo[2]

[1]Department of Pharmacobiology, CINVESTAV, México City 14330, México
[2]Escuela Superior de Medicina del IPN, México

Growing evidence indicates that antagonists of the 5-hydroxytryptamine (serotonin) receptor$_6$ (5-HT$_6$) improve memory and reverse amnesia, although the mechanisms involved are poorly understood. Hence, in this paper an attempt was made to summarize recent findings. Available evidence indicates that diverse 5-HT$_6$ receptor antagonists produce promnesic and/or antiamnesic effects in diverse conditions, including memory formation, age-related cognitive impairments, memory deficits in diseases such as schizophrenia, Parkinson, and Alzheimer's disease (AD). Notably, some 5-HT$_6$ receptor agonists seem to have promnesic and/or antiamnesic effects. At the present, it is unclear why 5-HT$_6$ receptor agonists and antagonists may facilitate memory or may reverse amnesia in some memory tasks. Certainly, 5-HT$_6$ drugs modulate memory, which are accompanied with neural changes. Likewise, memory, aging, and AD modify 5-HT$_6$ receptors and signaling cascades. Further investigation in different memory tasks, times, and amnesia models together with more complex control groups might provide further clues. Notably, human studies suggest a potential utility of 5-HT$_6$ receptor antagonists in mild-to-moderate AD patients. Even individuals with mild cognitive impairment (MCI) offer a great opportunity to test them.

INTERNATIONAL REVIEW OF
NEUROBIOLOGY, VOL. 96
DOI: 10.1016/B978-0-12-385902-0.00002-4

I. Introduction

Memory may be defined according to its content, in relation to time and its neurobiological basis: in the former case, as declarative/explicit or nondeclarative/implicit memory, and regarding time, as short-term (STM) or working, and long-term memory (LTM) (Davis and Squire, 1984; Izquierdo *et al.*, 1999, 2006); the latter depends on protein and mRNA synthesis. Notably, diverse brain areas and neurotransmission systems mediate memory systems, including serotonergic systems. Serotonin (5-hydroxytryptamine, 5-HT) was discovered about 50 years ago, and currently it still continues to generate interest as one of the most successful targets for therapeutic applications, ranging from depression, schizophrenia, anxiety to learning and memory disorders. Serotonin has been implicated in learning and memory previously (revised by Altman and Normile, 1988; Ögren, 1985), and recently this notion has gained wider acceptance and interest. For instance, a Medline search in 1999 for 5-HT and memory or 5-HT and learning produced 394/1512 articles, respectively; whereas, in a March (2007) search for serotonin and memory or 5-HT and learning or 5-HT and memory or serotonin and learning yielded 4113 publications. Thus, in the last few years more than 2500 or 3500 papers have appeared that directly or indirectly implicate 5-HT or its receptors in learning and memory in species ranging from humans to invertebrates, and to date (December 2010), this trend continues (5311 papers).

II. 5-HT Systems and Markers Related to Memory Systems

The identification of seven 5-HT receptor families (5-HT$_1$ to 5-HT$_7$) (Hannon and Hoyer, 2008; Hoyer *et al.*, 2002) and serotonin transporter (SERT; see e.g., Kalueff *et al.*, 2010) in mammalian species as well as drugs selective for these sites have allowed to dissect their participation in learning and memory (see e.g., Dayan and Huys, 2009; King *et al.*, 2008; Meneses, 1999, 2003; Pérez-García and Meneses, 2008a; Meneses *et al.*, 2011; Terry *et al.*, 2008). Actually, growing evidence indicates that 5-HT receptors and SERT are involved in normal, pathophysiological, and therapeutic aspects of learning and memory (Meneses, 1999; Meneses *et al.*, 2011; Schmitt *et al.*, 2006). For instance, disorders such as AD and schizophrenia contain an important component of dysfunctional memory and their etiology includes dysfunctions of cholinergic, glutamatergic, and serotonergic systems (Cassel, 2010; Li *et al.*, 2007; Meneses and Perez-Garcia, 2007; Pérez-García and Meneses, 2008a; Steckler and Sahgal, 1995; Stewart *et al.*, 2007; Terry *et al.*, 2008). The serotonergic systems have been also implicated in diseases with memory disorders such as depression, Down's syndrome, posttraumatic stress disorder, and so on.

5-HT pathways project to almost all brain areas (Jacobs and Azmitia, 1992; Steinbusch, 1981) and diverse 5-HT mechanisms might be useful in the treatment of learning and memory dysfunctions. Notably, aging, AD, and amnesia are associated with decrements in 5-HT markers such as the raphe nuclei, the uptake/transporter complex, and in the number of 5-$HT_{1A\text{-}1D}$, 5-$HT_{2A/2C}$, 5-HT_4, 5-HT_6, and 5-HT_7 receptors (for references, see Francis *et al.*, 2010; Meneses, 1999, 2003; Meneses and Perez-Garcia, 2007; Meneses *et al.*, 2011). Emerging evidence suggests that memory formation, amnesia, and promnesic and amnesic drugs modify serotonergic markers, including 5-HT receptors and SERT (see e.g., Belcher *et al.*, 2005; Garcia-Alloza *et al.*, 2004; Huerta-Rivas *et al.*, 2010; Lorke *et al.*, 2006; Marcos *et al.*, 2006, 2008; Meneses *et al.*, 2007; 2011; Pérez-García and Meneses, 2006, 2008a, 2009; Tellez *et al.*, 2010). Certainly, 5-HT_{1A}, 5-$HT_{2A/2C}$, 5-HT_3, 5-HT_4, 5-HT_6, and 5-HT_7 receptors have attracted more scientific interest regarding memory (see e.g., Bockaert *et al.*, 2008; Meneses, 1999, 2003; King *et al.*, 2006, 2008; Ögren *et al.*, 2008; Roth *et al.*, 2004; Terry *et al.*, 2008). Particularly, 5-HT_{1A}, 5-HT_4, 5-HT_6, and 5-HT_7 receptors have in common that their agonists and/or antagonists seem to have promnesic and/or antiamnesic effects (see e.g., Bockaert *et al.*, 2008; King *et al.*, 2008; Meneses, 1999, 2003; Ögren *et al.*, 2008; Roth *et al.*, 2004; Terry *et al.*, 2008; van Praag, 2004).

As 5-HT_6 receptors have attracted wide interest, reviewing more current pharmacological, physiological (Tassone *et al.*, 2010), behavioral, and molecular studies may be particularly insightful and timely in view of the apparently contradictory notion that either 5-HT_6 receptor agonists or antagonists might be useful in the treatment of learning and memory disorders. In this regard, probably a key issue to determine is whether 5-HT markers (e.g., receptor, reuptake sites, neural levels of serotonin) directly or indirectly contribute to the physiological and pharmacological basis of memory and its pathogenesis (Meneses and Perez-Garcia, 2007), or if they represent protective or adaptable mechanisms. Finally, it should be noted a recent review of the physiological effects associated to 5-HT_6 receptor (Tassone *et al.*, 2010), including the depolarizing response to 5-HT was reduced by the selective 5-HT_6 receptor antagonist SB258585. Moreover, application of 5-HT, in the presence of RS102221 plus SB269970 to selectively activate 5-HT_6 receptor, caused an increase in input resistance and induced an inward current.

III. Localization of the 5-HT_6 Receptor in Brain Areas Mediation Memory

Relative to the ontogenetic profiles of other 5-HT receptors, 5-HT_6 receptor mRNAs are expressed almost exclusively within the CNS and relatively early in brain development (see e.g., King *et al.*, 2008; Wilson and Terry, 2009), but are not expressed in 5-HT neurons (Woolley *et al.*, 2004). The distribution of 5-HT_6

receptor protein is in good agreement with the mRNA results in human brain and rats but in not mice (Liu and Robichaud, 2009). 5-HT_6 receptors are localized in brain areas involved in learning and memory processes (Zola-Morgan and Squire, 1993), including hippocampus, amygdala, striatum, and neocortex (King *et al.*, 2008, 2009; Meneses, 1999; Roberts *et al.*, 2002; Ward *et al.*, 1995; Woolley *et al.*, 2003). Specifically, the higher density of 5-HT_6 receptor is located in striatum, olfactory tubercles, and nucleus accumbens, with moderate levels in the cerebral cortex, hippocampus (CA1–CA3, dentate gyrus), hypothalamus, and amygdala (Gerard *et al.*, 1996, 1997; Woolley *et al.*, 2003). 5-HT_6 receptor modulates cholinergic, dopaminergic, glutamatergic, and serotonergic function (Bourson *et al.*, 1995; Dawson *et al.*, 2001; Hirst *et al.*, 2006; King *et al.*, 2008; Riemer *et al.*, 2003; Woolley *et al.*, 2004). Importantly, 5-HT_6 receptor is coupled to stimulatory G-proteins (Gs) and therefore stimulates cAMP signaling pathway (Ruat *et al.*, 1993) and other associated molecular mechanisms underlying learning and memory (see e.g., Kandel, 2001; Linden *et al.*, 2008; Sossin, 2008).

IV. 5-HT_6 Receptors Signaling

The human and mouse 5-HT_6 receptors are glycoproteins of 440 amino acids. In rats, the protein has 438 amino acids and all known homologs have seven transmembrane domains that form three cytoplasmic and extracellular loops. The mechanism of interaction between serotonin receptors and their G-proteins shows that in the case of the 5-HT_6 receptor, the third intracellular loop (iL3) and the C-terminal cytosolic (CT) region have fairly long stretches of amino acids, 55 and 104 residues, respectively. In addition, a sequence comparison indicated that iL2 and iL3 of the 5-HT_6 receptors have strong sequence homology to other serotonin receptors. These observations suggest that the iL2, iL3, or CT regions of the 5-HT_6 receptor may interact directly with G-proteins. 5-HT_6 receptors are positively coupled to the Gs protein, inducing cAMP production. In addition, recent *in vitro* evidence indicates a functional coupling of the carboxyl terminus of the human 5-HT_6 receptor to Fyn tyrosine kinase (see below) and 5-HT_6 receptor mediation of K^+ channel depolarization in rat striatal cholinergic interneurons has also been suggested. The depolarizing response to 5-HT seems to be reduced by the selective 5-HT_6 receptor antagonist SB-258585 (see also Tassone *et al.*, 2010) and activation of 5-HT_6 receptors induced an inward current likely involving phospholipase C and adenylyl cyclase (AC) (Bonsi *et al.*, 2007). It has been noted (Bockaert *et al.*, 2006) that 5-HT_6 receptor gene presents a splicing site that lies in the middle of the intracellular loop 3; two separate exons encode TM6 and TM7 of the 5-HT_6 receptors and a nonfunctional splice variant has been described for this receptor. Moreover, studies performed mainly in heterologous cells transfected with both native and mutant forms of serotonergic receptors (see Bockaert *et al.*, 2006), 5-HT$_1$, 5-HT$_2$, 5-HT$_4$, 5-HT$_6$, and 5-HT$_7$, have demonstrated that they can exist in a conformation that mimics an active state of the receptor in the absence of agonist, thus revealing that these receptors display constitutive activity. Certainly, spontaneous signaling remains unreported for human 5-HT$_6$ receptors, although mutations in intracellular loop 3 (which interacts with Gs) render them constitutively active. For instance, in native mouse, but not rats (Fone, 2008), 5-HT$_6$ receptors display constitutive activity for coupling to cAMP (see Millan *et al.*, 2008). As already mentioned 5-HT$_6$ receptors recruit Gs and AC, but their cellular pharmacology was recently enlivened by the finding that their activation triggers phosphorylation of Fyn, a member of the Src family of nonreceptor tyrosine kinases. Fyn, in turn, activates the key intracellular modulator, extracellular regulated kinase (ERK)1/2, via a classical Ras–Raf1–MEK (mitogen-activated protein kinase) cascade (Yun *et al.*, 2007; see also Millan *et al.*, 2008). Reciprocally, in a form of positive feedback, binding of Fyn to the C terminus of 5-HT$_6$ receptors increases their cell-surface expression. Notably, Tau is a microtubule-associated protein and a main component of neurofibrillary tangles, one of the pathologic hallmarks of AD. Historically, most of the tau phosphorylation sites that have been characterized are serine and threonine residues; nevertheless, recent reports state that tau can be phosphorylated at tyrosine residues by kinases including Fyn (see Tremblay *et al.*, 2010). This is important since, in general terms (Millan *et al.*, 2008), a specific 5-HT-receptor subtype influences several G-proteins and downstream messengers (divergence). Likewise, multiple 5-HT receptors collectively impact (convergence) individual signals such as AC and ERK, in addition to transcription factors such as cAMP response-element-binding protein (CREB). Thus, underscoring prospects that drugs can directly affect intracellular signals, agents blocking the interaction between 5-HT$_6$ receptors and Gs have been recently unveiled (see Millan *et al.*, 2008). Finally, the manipulation of Jab1 expression using Jab1 small interference RNA decreased 5-HT$_6$ receptor-mediated activity and their cell membrane expression (Yun *et al.*, 2010). The activation of 5-HT$_6$ receptors induced the translocation of Jab1 into the nucleus and increased c-Jun phosphorylation and the interaction between Jab1 and c-Jun (Yun *et al.*, 2010), suggesting even another signal transduction pathway for these receptors.

V. Pharmacology of the 5-HT$_6$ Receptor and Memory

Investigation of 5-HT systems involvement on memory, particularly 5-HT$_6$ receptors, has been significantly enhanced by the identification of selective antagonists, and more recently, agonists (see e.g., King *et al.*, 2008; Wilson and Terry, 2009) as well as the investigation at the postreceptor level (see e.g., Hirano *et al.*, 2009; Huerta-Rivas *et al.*, 2007, 2010; Marcos *et al.*, 2008, 2010; Mitchell *et al.*,

2007, 2009; Pérez-García and Meneses, 2008a, 2008b). Certainly, mice lacking the 5-HT$_6$ receptor presented neither gross anatomical or behavioral abnormalities nor obvious changes in microscopic brain morphology, and their performance in rotarod, open field, and novel object testing paradigms revealed no differences compared with wild-type animals (for references, see Meneses, 2001b). These results are in line with the evidence that mice express very low levels of 5-HT$_6$ receptors (Hirst *et al.*, 2003). Notably, the interest about the role of 5-HT$_6$ receptors in memory has been growing. For instance, several important reviews have been published (Fone, 2008; Geldenhuys and Van der Schyf, 2009; Holenz *et al.*, 2006; Johnson *et al.*, 2008; Liu and Robichaud, 2009; Mitchell and Neumaier, 2005; Rossé and Schaffhauser, 2010; Russell and Dias, 2002; Terry *et al.*, 2008; Upton *et al.*, 2008; Wilson and Terry, 2009; Woolley *et al.*, 2004).

The first indirect evidence of 5-HT$_6$ receptor involvement in memory was provided by using antisense oligonucleotides. Within a few years, 5-HT$_6$ receptor antagonists showed that these drugs produced promnesic and/or antiamnesic effects in a number of memory tasks, including water maze, passive avoidance, autoshaping, fear conditioning, novel object recognition (NOD), and social memory (earlier reviews, see Branchek and Blackburn, 2000; Meneses, 2001b; Sleight *et al.*, 1998). Certainly, discrepant results appeared; for instance, Russell and Dias (2002), using the Morris water maze and both 5-HT$_6$ receptor antagonists Ro 04-6790 and SB-271046, failed to replicate the finding of an improved memory. In addition, no changes in autoshaping, Morris water maze, or fear conditioning were reported by Lindner *et al.* (2003). Although the discrepant results have not been completely clarified, some evidence suggests methodological differences (Pérez-García and Meneses, 2005, see below). Probably, the more important issue of the discrepant results has been the development of potent and selective 5-HT$_6$ receptor antagonists with improved pharmacokinetic parameters (e.g., brain penetration, oral bioavailability, plasma half-life), which has been a decisive step in this regard (see Johnson *et al.*, 2008; Wilson and Terry, 2009).

VI. 5-HT$_6$ Receptor Antagonists and Agonists on Memory Tasks

During several years 5-HT$_6$ receptor antagonists were the principal tools in the study of the role of these receptors on memory. Nevertheless, it has become clear recently that the 5-HT$_6$ receptor agonists produce promnesic effects on some memory tasks, for example, NOD (see below). Except for the discrepancies mentioned above (Lindner *et al.*, 2003; Russell and Dias, 2002), there is a general agreement that 5-HT$_6$ receptor antagonists enhance memory and/or have antiamnesic effects (for review, see Fone, 2008; Meneses, 2001a; Woolley *et al.*, 2004) in diverse available memory tasks (Lynch, 2004; Meneses and Perez-Garcia, 2007;

Myhrer, 2003; Peele and Vincent, 1989), including water maze (e.g., Hirst *et al.*, 2006), inhibitory avoidance (e.g., Foley *et al.*, 2004; Riemer *et al.*, 2003), autoshaping (e.g., Meneses, 2001a, 2001b; Pérez-García and Meneses, 2005; Schreiber *et al.*, 2007), and novel object discrimination (e.g., Fone, 2006; see also Mitchell and Neumaier, 2005; Schreiber *et al.*, 2007; Woolley *et al.*, 2004), thus confirming the promnesic and/or antiamnesic effects of diverse 5-HT_6 receptor antagonists. When comparing them and/or protocols of training, it would be logical to observe different outcomes than, for example, associative Pavlovian/instrumental autoshaping versus spatial memory (e.g., water maze) or aversive paradigms such as inhibitory avoidance. Doubtless as mentioned above, the development of potent and selective 5-HT_6 receptor antagonists has been crucial in the clarification of the role of 5-HT_6 receptors in memory. Thus, reinforcing the notion that antagonists for these receptors facilitate memory formation, reverse age-related impairments, and/or antiamnesic effects on pharmacological and genetic models, available evidence includes normal memory, aged animals, amnesia induced by scopolamine (cholinergic muscarinic antagonist), dizocilpine or phencyclidine (both glutamatergic antagonists), apomorphine (dopaminergic agonist), tryptophan depletion (precursor of serotonin), and striatal overexpression of 5-HT_6 receptors (King *et al.*, 2008; Mitchell *et al.*, 2007; Wilson and Terry, 2009). Notably, the cognitive enhancing effects of Ro 04-6790 (which poorly penetrates the brain, King *et al.*, 2008) in the NOD task appear to be mediated via brain regions whose serotonergic innervation is lost following the administration of the serotonergic neurotoxin 5,7-DHT into the lateral ventricles (LV) or medial, but not dorsal, raphe nuclei depletion (King *et al.*, 2009). According to these authors, the ability of 5-HT depletion to prevent the effect of Ro 04-6790, without altering basal NOD performance under vehicle conditions, confirms that the procognitive effects of Ro 04-6790 are due to blockade of tonic serotonergic neuronal activity. Thus, basal activation of the 5-HT_6 receptors by endogenous 5-HT normally inhibits cognitive processes (King *et al.*, 2008). While Ro 04-6790 had no effect on basal cAMP *in vitro* production, King *et al.* (2009) suggest that it is neither an agonist nor an inverse agonist. Moreover, while 5-HT depletion with *p*-chloroamphetamine did not alter performance on autoshaping task (a memory task that probably represents situations of self-taught), the three 5-HT_6 receptor antagonists Ro 04-6790, SB-399885 (which is over 200-fold more selective for the 5-HT_6 receptor and a potent competitive antagonist; Hirst *et al.*, 2006), and SB-357134 had no effects on STM but facilitated LTM, suggesting that a "5-HT tone via 5-HT_6 receptors" might be responsible that LTM works in parallel manner (Meneses, 2007b). Notably, as memory was progressive in the Pavlovian/instrumental autoshaping task, *ex vivo* mRNA expression of 5-HT_6 receptors was reduced in the prefrontal cortex (PFC), hippocampus raphe nuclei (Huerta-Rivas *et al.*, 2010). In addition, SB-399885 improved LTM at 48 h, but the muscarinic receptor antagonist scopolamine or the noncompetitive NMDA receptor antagonist dizocilpine impaired it at 24 h. Autoshaping training plus SB-399885 treatment increased 5-HT$_6$ receptor mRNA expression in (maximum increase) PFC and striatum, at 24 or 48 h. The scopolamine- or dizocilpine-induced amnesia suppressed or did not modify 5-HT$_6$ receptors mRNA expression. The SB-399885 plus scopolamine or dizocilpine reestablished memory and 5-HT$_6$ receptor mRNA expression.

Notwithstanding the basal level in the control trained animals showed a very modest level of autoshaped responses, which had been previously reported (see e.g., Meneses, 2003; Pérez-García *et al.*, 2006; also Fone, 2008), they tended to show 5-HT$_6$ receptor mRNA downregulation in PFC, striatum (STR), and hippocampus (HIP) at 1.5, 24, and 48 h; mainly in the later time in PFC and striatum. In contrast, memory and SB-399885 slightly tend to upregulate 5-HT$_6$ receptor mRNA expression in PFC with respect to untrained SB-399885-treated rats, showing the maximum increase in PFC (at 24 h) and striatum (at 48 h). Also improvement memory at 24 and 48 h upregulated 5-HT$_6$ mRNA expression in STR, with respect to trained control group. Scopolamine or dizocilpine in untrained animals did not modify 5-HT$_6$ mRNA basal expression. Nevertheless, the scopolamine amnesic effect was accompanied by a significant 5-HT$_6$ mRNA downregulation in the three areas explored. Dizocilpine amnesic effect was associated with slight decreases in 5-HT$_6$ mRNA. SB-399885 was able to reestablish memory and, in part, upregulate 5-HT$_6$ receptor mRNA expression in animals treated with scopolamine and slightly in those treated with dizocilpine. These data are related to several recent demonstrations (see Cammarota *et al.*, 2008; Kandel, 2001; McGaugh, 2006; Radley *et al.*, 2007; Sweatt, 2009; Ye and Carew, 2010), particularly in the context of memory and neuronal markers, with focus on (1) What is the effect of training on expression of the receptor mRNA? (2) What is the effect of memory-modulating drugs on the receptor? (3) Can the effect of a memory-impairing drug at the receptor level be reversed by blocking it? The answers to these questions seem to be encouraged. For instance, in an autoradiographic study of 5-HT$_6$ receptors at protein level (Meneses *et al.*, 2008), SB-399885 improved memory consolidation and decreased 5-HT$_6$ receptor expression in 15 out 17 brain areas. The scopolamine- or dizocilpine-induced amnesia decreased 5-HT$_6$ receptor expression in nine different brain areas and increased it in CA3 hippocampus or other eight areas, respectively, including brain areas thought to be in charge of procedural memory such as basal ganglia (i.e., nucleus accumbens, caudate putamen). Certainly, relative to control animals amnesic groups showed diminished (scopolamine) or augmented (dizocilpine) 5-HT$_6$ receptor expression. The SB-399885 improved memory displayed an intermediate expression in these same brain regions. A similar intermediate expression occurs with regard to amygdala, septum, and some cortical areas in charge of explicit memory storage. However, relative to the control animals amnesic and SB-399885 groups in the hippocampus, region where explicit memory is formed, showed a complex 5-HT$_6$ receptor expression. It should be noted that, in untrained animals, neither scopolamine nor dizocilpine modified the basal expression of 5-HT$_6$ receptor mRNA. A similar finding was reported with dizocilpine (0.3 mg/kg for 7 days), which did not modify basal expression in STR (Healy and Meador-Woodruff, 1999). Nonetheless, trained rats showed amnesia at 24 h and downregulated 5-HT$_6$ receptor mRNA expression in the three structures analyzed, almost undetectable (scopolamine); in contrast, upregulated 5-HT$_6$ receptor mRNA expression in striatum seemed to be associated with the dizocilpine amnesia. Certainly these differences in 5-HT$_6$ receptor mRNA expression between scopolamine and dizocilpine might be related to the neural circuit recruited by each amnesic drug. On the other hand, trained control rats presented a slightly decreased expression in PFC, STR, and HIP relative to untrained control animals, suggesting that in basal conditions (i.e., no learning and no treatment), 5-HT$_6$ receptor mRNA was more abundant and that learning and memory slightly downregulated it. This was particularly observable in PFC at 24–48 h and in STR at 1.5 and 48 h; even it was almost immediately observable after the training session (Huerta-Rivas *et al.*, 2010). Relative to untrained control animals, SB-399885 treatment in untrained group downregulated 5-HT$_6$ receptor mRNA expression in PFC without changes in HIP or STR. Once again, notwithstanding the modest memory improvement (or amnesic changes) observed on autoshaping task, 5-HT$_6$ receptor mRNA upregulation in (slightly) PFC and (significantly) in STR induced by SB-399885 and the facilitation of memory might be interpreted as a demand for further receptor synthesis. Interestingly, although hippocampal 5-HT$_6$ receptor mRNA expression was not modified during memory consolidation (Huerta-Rivas *et al.*, 2010), it was importantly decreased in the scopolamine- or (slightly) in the dizocilpine-induced amnesia. SB-399885 reestablished both memory and 5-HT$_6$ receptor mRNA expression (Huerta-Rivas *et al.*, 2007, 2010). Notably, these data parallel evidence reported at 5-HT$_6$ receptor protein (see above, Meneses *et al.*, 2007) and preventing intrahippocampal protein or mRNA synthesis impaired memory (Meneses, 2007a). Hence, 5-HT receptors might represent a mechanism for selection of the moment and information type, which will pass from a labile state to a long-term memory or to be forgotten (Meneses, 2007a, 2007b). This is an important finding, suggesting that for an improve memory is necessary to reduced 5-HT$_6$ receptors expression, which facilitated processing of information. Nevertheless, if 5-HT$_6$ receptor expression suffers an extreme reduction (or overexpression), memory likely will be adversely affected. As occurred with the increased 5-HT$_6$ receptor expression in caudate putamen of the dizocilpine-treated amnesic (see above, Huerta-Rivas *et al.*, 2007, 2010) rats and evidence reported by Mitchell *et al.* (2007) of 5-HT$_6$ receptor overexpression in the dorsal striatum showing poor memory of 12 sessions, only during three to seven sessions, in instrumental learning, but not in the water maze, and the memory impairment was antagonized by a 5-HT$_6$ antagonist, or prevented if the overexpression was induced once the animals had learned the task

(Mitchell *et al.*, 2007). Certainly, although reducing 5-HT receptor expression in some brain areas seems to be necessary for memory formation, a subtle modulation is important (Huerta-Rivas *et al.*, 2007, 2010); even overexpression suggests a specific time window (Mitchell *et al.*, 2007). Thus, the prevention of the pharmacologically induced amnesia in adult animals might be related to restoration (scopolamine) or maintenance (dizocilpine) of 5-HT$_6$ receptor mRNA expression at basal levels. Since the possible link between 5-HT$_6$ receptor and Ras family proteins, it should be noted that the fact that these proteins are highly dynamic may allow them to integrate signals across multiple learning experiences, and determine the optimal spacing between learning trials during incremental learning (Ye and Carew, 2010). Indeed, the dynamic profiles of ERK activity, a canonical downstream element of Ras family proteins, significantly constrain the temporal window between learning trials that are permissive for memory formation (see Ye and Carew, 2010).

An important consideration is in order, namely, the interaction among brain areas, neurotransmitters systems, drugs administration, and cognitive and behavioral demand of learning tasks and extent of training. For instance, memory formation requires hippocampus in behavioral tasks such as water maze (e.g., Kandel, 2001), passive avoidance (Izquierdo *et al.*, 1999), and Pavlovian/instrumental autoshaping (Meneses, 2003; Meneses *et al.*, 2009; Tellez *et al.*, 2010). In contrast, overtrained animals engage more striatum and less hippocampus in autoshaping (Meneses, 2003; Tellez *et al.*, 2010) and passive avoidance (Izquierdo *et al.*, 2006). While explicit or declarative memory has been related to hippocampus, implicit or nondeclarative memory has been related to striatum (e.g., Adamantidis and de Lecea, 2009; Kandel, 2001). Nevertheless, multiple memory mechanisms can work in tandem to support performance on an implicit memory task and even additional contribution of explicit memory can be observed in neurologically healthy individuals (Koenig *et al.*, 2008) or during acquisition of an otherwise implicit learning task (see also Meneses *et al.*, 2011). This implicates that during the early stage (i.e., learning) of implicit memory it does not depend on striatum; nevertheless, as training and time go through implicit memory eventually depends on striatum (Izquierdo *et al.*, 2006). Thus, it seems to be appropriate to reconsider the notion that, regardless of time of training and testing explicit and implicit memory are mediated by hippocampus and striatum, respectively. These considerations are quiet important in the context of neural markers. The case of 5-HT$_6$ receptors provides an excellent and timely instance of these issues. As already mentioned, rats overexpressing dorsomedial striatum, but not dorsocentral striatum, 5-HT$_6$ receptors showed impaired performance in a simple operant learning task (a striatum-dependent learning model), but not in the hippocampus-dependent water maze task (Mitchell *et al.*, 2007). Notably, this impairment effect was appreciable at third instrumental testing session or the second extinction session on performance of previously acquired instrumental conditioning (Mitchell *et al.*, 2007). In contrast, Pavlovian/instrumental autoshaping learning task might be conceptualized as an instance of systems processing stimulus–stimulus, stimulus–response, and stimulus–reinforcer learning (Meneses, 2003), which requires brain areas such as dentate gyrus, hippocampal CA1, basolateral amygdaloid nucleus, and PFC (Pérez-García and Meneses, 2008a). Thus, on autoshaping task, memory formation requires a serotonergic tone (at least) via 5-HT_6 receptors, suppressing their expression. Nevertheless, under amnesic conditions they are completely suppressed or slightly reduced; in contrast, when SB-399885 improves memory or amnesia is reversed the expression of 5HT_6 receptors is increased or reestablished, respectively. Moreover, memory formation on the water maze (MWM) downregulated 5-HT_6 receptor protein and mRNA receptor expression and the administration of the selective 5-HT_6 receptor antagonist SB-271046 induced an increase in pCREB1 levels while CREB2 levels were significantly reduced (Marcos *et al.*, 2010). However, although SB-271046 was able to improve retention in the MWM, no further changes in pCREB1 or CREB2 levels were observed. The MWM alone significantly increased pERK1/2 levels and further increases were seen when treating with SB-271046 during the MWM. According to Marcos *et al.* (2010) these results suggest that, in the hippocampus, biochemical pathways associated with pERK1/2 expression, and not with the CREB family of transcription factors, seem to be related to the cognitive-enhancing properties of 5-HT_6 receptor antagonists. In addition, repeated treatment with the 5-HT_6 antagonist RO4368554 (5.0 mg/kg) improved NOD and social discrimination without changing plasticity-associated proteins Ki-67 and PCNA (Mitchell *et al.*, 2009). These data suggest that diverse molecular mechanisms may be associated with promnesic or antiamnesic effects of 5-HT_6 receptor antagonists. Hence, the 5-HT_6 antagonists procognitive effects may be due to an indirect regulation of cholinergic (Bourson *et al.*, 1995), glutamatergic (King *et al.*, 2004), or even serotonergic (Hirst *et al.*, 2006; Lieben *et al.*, 2005) transmission through GABA release in PFC, hippocampus, and striatum, important brain areas for memory and associative learning (Meneses, 2003; see also Packard and Knowlton, 2002). In addition, serotonin clearly exerts a tonic effect upon 5-HT_6 receptor as shown by 5-HT_6 receptor antagonists' procognitive or antiamnesic effect. In mechanistic terms, an interesting possibility is that 5-HT_6 receptor activates the extracellular signal regulated kinase1/2 (ERK2) via a Fyn-dependent pathway, a member of the Src family of nonreceptor protein tyrosine kinases (for references, see Yun *et al.*, 2007). Fyn is involved in AD through the modulation of microtubule-associated tau and amyloid; Fyn deficiency results in defective neuronal and spatial memory.

Regarding 5-HT_6 receptor agonists, it was recently reported that 5-HT_6 receptor agonists such as EMD-386088 (at 5.0, but not 1.0 and 10.0 mg/kg) impaired STM and LTM (24 h) in autoshaping (Meneses *et al.*, 2008) task or WAY-466 (Schechter *et al.*, 2004) impaired memory in a social recognition task (Loiseau *et al.*, 2008). Nonetheless, if animals treated with either scopolamine or dizocilpine were given EMD-386088 (1.0 mg/kg), the induced amnesic

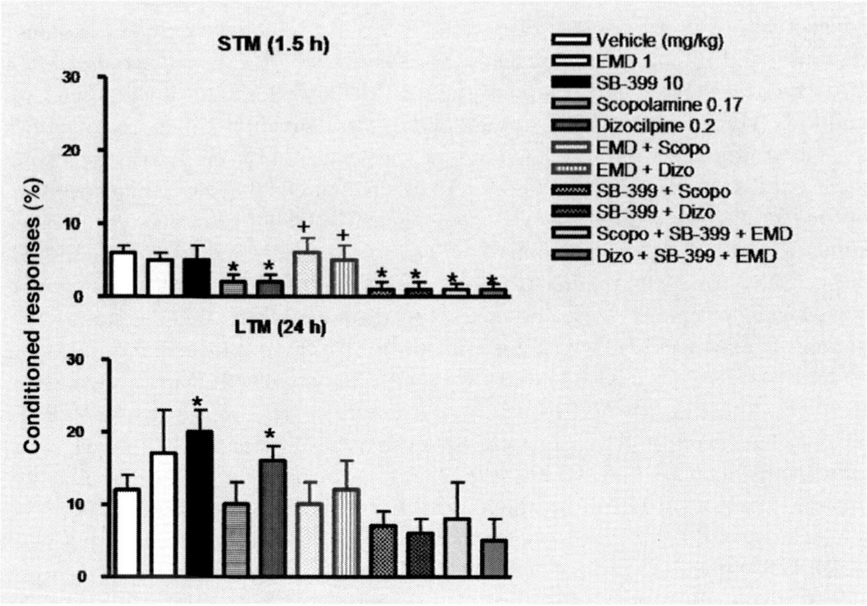

FIG. 1. Antiamnesic effects of the 5-HT_6 receptor agonist (EMD) and antagonist (SB-399885). Trained animals received saline (control vehicle), scopolamine (0.17 mg/kg i.p.), dizocilpine (0.2 mg/kg i.p.), EMD (1.0 mg/kg s.c.), and SB-399885 (10.0 mg/kg p.o.), and were tested in the autoshaping task at 1.5 (STM) and 24 and 48 h (LTM). In addition, in the same times trained animals were administered with vehicle or the combination of SB-399885 + scopolamine, SB-399885 + dizocilpine, EMD + scopolamine or EMD + dizocilpine, SB-399885 + EMD + scopolamine or SB-399885 + EMD + dizocilpine and tested in the autoshaping task for STM and LTM. Data (CR%) shown correspond to the STM and LTM (24 h). While EMD and SB-399885 had no effects on STM, either scopolamine or dizocilpine impaired it. The combination SB-399885 + EMD + scopolamine or SB-399885 + EMD + dizocilpine showed STM-impairment, showing that 5-HT_6 receptors were involved in antiamnesic effects of EMD. LTM (24 h) was unaltered or improved by SB-399885. Agonists or antagonists versus control $*p < 0.05$ + Tukey or +combinations versus agonists or antagonists. (For interpretation of the references to color in this figure legend, the reader is referred to the Web version of this chapter.)

effects on autoshaping task were reversed, and SB-399885 antagonized such as effects (Fig. 1), thus confirming the role of 5-HT_6 receptors. Notably, in the NOD paradigm the 5-HT_6 receptor agonist R-13c or E-6801 improved memory in a 4-h intertrial interval (for review, see Fone 2008). More recently Kendall *et al.* (2010) reported that the 5-HT_6 receptor agonists, E-6801 (1.25–10.0 mg/kg) and EMD-386088 (5.0–10.0 mg/kg), and antagonists, SB-271046 and Ro 04-6790 (5.0 and 10.0 mg/kg), along with donepezil (0.1–3.0 mg/kg) or memantine (5.0–20.0 mg/kg) all produced significant and mostly dose-dependent improved memory. Furthermore, subeffective doses of E-6801 (1.0 mg/kg) when coadministered with SB-271046 (3.0 mg/kg), donepezil (0.1 mg/kg), or memantine (5.0 mg/kg), and EMD-386088 (2.0 mg/kg) coadministered with SB-271046

(3.0 mg/kg) also significantly enhanced NOD memory. Additionally, using a 1-min intertrial delay, E-6801 (2.5 and 5.0 mg/kg) was as effective as donepezil (0.3 and 1.0 mg/kg) in reversing a scopolamine-induced (0.5 mg/kg) impairment in NOD. According to Kendall *et al.* (2010), this was the first study to demonstrate that E-6801, a potent 5-HT_6 receptor agonist, improves recognition memory by combined modulation of cholinergic and glutamatergic neurotransmission. Similarly, the 5-HT_6 receptor agonist WAY-181187 enhanced the extradimensional set shift in the rat attentional set-shifting paradigm (Burnham *et al.*, 2007, 2010); importantly, 5-HT_6 receptor antagonists improve performance in this same task (Hatcher *et al.*, 2005). However, in the social recognition paradigm, WAY-181187 given systemically or by discrete bilateral injections into the frontal cortex impaired performance, an effect abolished by the antagonists SB-271046 or SB-258585 (Loiseau *et al.*, 2008). Notably, in the long-term potentiation (LTP, a physiological model of memory) WAY-181187 activation of hippocampal CA1 area had no effect on baseline synaptic transmission, but attenuated it over a narrow dose range and this effect was dose-dependently blocked by SB-399885 (West *et al.*, 2009; see also Wilson and Terry, 2009). WAY-181187 also increased the frequency of spontaneous GABA release in area CA1 and was prevented by the selective 5-HT_6 antagonist SB-399885. In addition, systemic administration of 5-HT_6 antagonists increases the release of acetylcholine and glutamate in the frontal cortex and dorsal hippocampus. In contrast, the selective 5-HT_6 agonist, WAY-181187, can elicit robust increases in extracellular levels of GABA. Recent evidence indicates that EMD and SB-388995 improved (at 48 h) LTM in autoshaping task and modulated in PFC, hippocampus, striatum, and raphe nuclei cAMP production relative to the trained control group (Table I); thus, stimulation of 5-HT_6 receptors was associated with cAMP increases in all four areas while their blockade was accompanied by modest increases or even decreases (hippocampus). Notably, density and 5-HT levels were significantly decreased in a cohort of AD patients prospectively assessed for cognitive/behavioral symptoms (Marcos *et al.*, 2008). Also, cAMP formation after stimulation with the selective 5-HT_6 receptor agonist E-6801 was significantly decreased compared with controls. In addition, the ratio cAMP formation after stimulation with E-6801/5-HT_6 receptor density was significantly lower in AD compared with control individuals; 5-HT_6 receptor activation was significantly lower in AD females. These authors related the psychosis factor as the best predictor of reduced 5-HT levels or adenylate cyclase activity after E-6801 stimulation, the former result being due to females. Finally, administration of the acute 5-HT_6 receptor agonist LY-586713 caused BDNF and Arc expression in cortical and hippocampal regions, suggesting that LY-586713 has potential effects on neuronal plasticity, such as effects were reversed by SB-271046, which alone increased Arc expression in these regions (de Foubert *et al.*, 2007).

Certainly why and how 5-HT_6 receptor agonists and antagonists may facilitate memory or reverse amnesia in some memory tasks (see Fone, 2008; Huerta-Rivas

Table I
COMPARISON AMONG BRAIN AREAS AND cAMP PRODUCTION[a] DURING LTM + EMD OR SB-399885 VERSUS SALINE TRAINED GROUP.

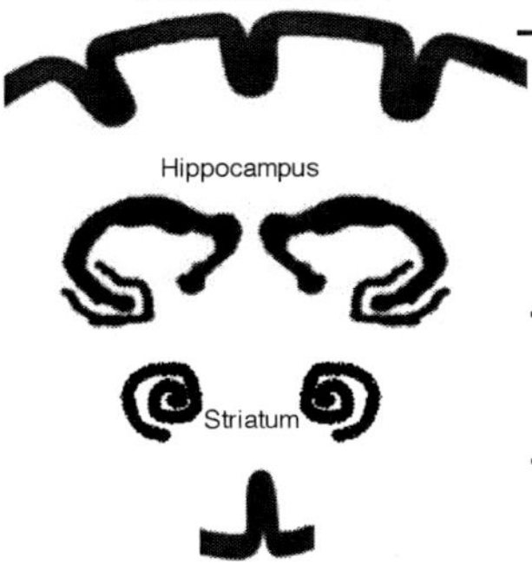

	Vehicle	5-HT6 Agonist	5-HT6 Antagonist
Prefrontal cortex	100	203 ±19[a]	199 ± 2[a]
Hippocampus	100	210 ±15[a]	71 ±6[a]
Striatum	100	200 ± 19[a]	171 ± 1[a]
Raphe nuclei	100	173 ± 5[a]	102 ± 2

[a]CONCENTRATION OF cAMP pmol PER mg/tissue.

et al., 2010; Upton *et al.*, 2008) left open very interesting possibilities. Doubtless, further experiments are necessary to clarify the role of 5-HT$_6$ receptors during memory formation, by testing diverse compounds, neural markers, and behavioral tasks. The development of effective treatments for memory alterations is, nonetheless, limited by the absence of reliable markers to indicate and/predict efficacy. This is important in the context that human studies suggest a potential use of 5-HT$_6$ receptor agonists and or antagonists, (e.g., Geldenhuys and Van der Schyf, 2009; Reid *et al.*, 2010; Upton *et al.*, 2008). This premise has been translated into the clinical efficacy of some 5-HT$_6$ receptor antagonists, including SB-742457, GSK-773812, SYB-114, SAM-531, and SUVN-502 in mild-to-moderate AD patients (Liu and Robichaud, 2009; Rossé and Schaffhauser, 2010). Even individuals with MCI offer a great opportunity to test 5-HT$_6$ receptor antagonists. Ivachtchenko *et al.* (2010) have recently published a review of patents involving 5-HT$_6$ receptor, suggesting that its functions might be mediated by inverse agonists.

VII. Conclusion and Final Comments

The role of 5-HT$_6$ receptors on memory has becomes a major area of scientific interest. Clearly, available data show the interaction among 5-HT$_6$

receptor protein and mRNA expression, memory, amnesia, and prevention of the amnesia. Further interest is provided by the evidence that both 5-HT_6 receptor agonists and antagonists may have promnesic and/or antiamnesic effects, in conditions covering memory formation, age-related cognitive impairments, and memory deficits in diseases such as schizophrenia, Parkinson, and AD. Certainly, this situation is not new regarding serotonergic receptors (e.g., 5-HT_{1A} receptor agonists and antagonists). One way to address such questions is by testing 5-HT_6 receptor agonists and antagonists in memory tasks and to analyze changes in neurobiological markers (e.g., 5-HT_6 receptor protein or mRNA expression, signaling cascades, etc.), in animals under memory formation or amnesia and drug administration. Importantly, the identification of reliable neural markers is fundamental for the understanding of memory mechanisms, its alterations, and potential treatment. The memory and molecular changes might represent promising steps, mainly in the light of pharmacological magnetic resonance imaging (phMRI; Martin and Sibson, 2008), which offers potential novel insights into the functioning of neurotransmitter systems and drug action in the CNS. Such studies might be useful to provide markers of the neuropharmacological modulation of neuronal activity across the whole brain with spatial and temporal specificity. Further investigation using different memory tasks, times, and amnesia models might provide important clues.

Acknowledgments

We thank Sofia Meneses-Goytia for revised language and Roberto Gonzalez for his expert assistance. Finally, our special thanks to anonymous referees who made important comments and valuable suggestions to previous papers. This work was supported in part by CONACYT grant 80060.

References

Adamantidis, A., and de Lecea, L. (2009). A role for melanin-concentrating hormone in learning and memory. *Peptides* **30**, 2066–2070.

Altman, H.J., and Normile, H.J. (1988). What is the nature of the role of the serotonergic nervous system in learning and memory: prospects for development of an effective treatment strategy for senile dementia. *Neurobiol. Aging* **9**, 627–638.

Belcher, A.M., O'Dell, S.J., and Marshall, J.F. (2005). Impaired object recognition memory following methamphetamine, but not *p*-chloroamphetamine- or D-amphetamine-induced neurotoxicity. *Neuropsychopharmacology* **30**(11); 2026–2034.

Bockaert, J., Claeysen, S., Bécamel, C., Dumuis, A., and Marin, P. (2006). Neuronal 5-HT metabotropic receptors: fine-tuning of their structure, signaling, and roles in synaptic modulation. *Cell Tissue Res.* **326**, 553–572.

Bockaert, J., Claeysen, S., Compan, V., and Dumuis, A. (2008). 5-HT_4 receptors: history, molecular pharmacology and brain functions. *Neuropharmacology* **55**, 922–931.

Bonsi, P., Cuomo, D., Ding, J., Sciamanna, G., Ulrich, S., Tscherter, A., Bernardi, G., Surmeier, D.J., and Pisani, A. (2007). Endogenous serotonin excites striatal cholinergic interneurons via the activation of 5-HT_{2C}, 5-HT_6, and 5-HT_7 serotonin receptors: implications for extrapyramidal side effects of serotonin reuptake inhibitors. *Neuropsychopharmacology* **32**, 1840–1854.

Bourson, A., Borroni, E., Austin, R.H., Monsma Jr., F.J., and Sleight, A.J. (1995). Determination of the role of the 5-ht_6 receptor in the rat brain: a study using antisense oligonucleotides. *J. Pharmacol. Exp. Ther.* **274**, 173–180.

Branchek, T.A., and Blackburn, T.P. (2000). 5-ht_6 receptors as emerging targets for drug discovery. *Annu. Rev. Pharmacol. Toxicol.* **40**, 319–334.

Burnham, K.E., Baxter, M.G., Bainton, J.R., Southam, E., Dawson, L.A., Bannerman, D.M., and Sharp, T. (2010). Activation of 5-HT_6 receptors facilitates attentional set shifting. *Psychopharmacology (Berlin)* **208**, 13–21.

Burnham, K. E., Baxter, M. G., Dawson, L. A., Southam, E., Sharp, T., and Bannerman, D. M. (2007). Effect of the 5-HT_6 receptor agonist WAY181187 on prefrontal cortical function in the rat. Neuroscience Abstract 741.28.AAA1.

Cammarota, M., Bevilaqua, L.R., Medina, J.H., and Izquierdo, I. (2008). ERK1/2 and CaMKII-mediated events in memory formation: is 5HT regulation involved? *Behav. Brain Res.* **195**, 120–128.

Cassel, J.C. (2010). Experimental studies on the role(s) of serotonin in learning and memory functions. Müller, C.P., Jacobs, B.L. (Eds.), *Handbook of the Behavioral Neurobiology of Serotonin.* **Vol. 21**, Academic Press, Amsterdam, pp. 429–447.

Davis, H.P., and Squire, L. (1984). The pharmacology of memory: a review. *Psychol. Bull.* **96**(3); 518–559.

Dawson, L.A., Nguyen, H.Q., and Li, P. (2001). The 5-HT_6 receptor antagonist SB-271046 selectively enhances excitatory neurotransmission in the rat frontal cortex and hippocampus. *Neuropsychopharmacology* **25**, 662–668.

Dayan, P., and Huys, Q.J. (2009). Serotonin in affective control. *Annu. Rev. Neurosci.* **32**, 95–126.

de Foubert, G., O'Neill, M.J., and Zetterström, T.S. (2007). Acute onset by 5-HT_6-receptor activation on rat brain brain-derived neurotrophic factor and activity-regulated cytoskeletal-associated protein mRNA expression. *Neuroscience* **147**(3); 778–785.

Francis, P.T., Ramírez, M.J., and Lai, M.K. (2010). Neurochemical basis for symptomatic treatment of Alzheimer's disease. *Neuropharmacology* **59**, 221–229.

Foley, A.G., Murphy, K.J., Hirst, W.D., Gallagher, H.C., Hagan, J.J., Upton, N., Walsh, F.S., and Regan, C.M. (2004). The 5-HT_6 receptor antagonist SB-271046 reverses scopolamine-disrupted consolidation of an inhibitory avoidance task and ameliorates spatial task deficits in aged rats. *Neuropsychopharmacology* **29**, 93–100.

Fone, K.C. (2006). Selective 5-HT_6 compounds as a novel approach to the treatment of Alzheimer disease. *J. Pharmacol. Sci.* **101**(Suppl. 1); 53.

Fone, K.C. (2008). An update on the role of the 5-hydroxytryptamine$_6$ receptor in cognitive function. *Neuropharmacology* **55**, 1015–1022.

Garcia-Alloza, M., Hirst, W.D., Chen, C.P., Lasheras, B., Francis, P.T., and Ramírez, M.J. (2004). Differential involvement of 5-$HT_{1B/1D}$ and 5-HT_6 receptors in cognitive and non-cognitive symptoms in Alzheimer's disease. *Neuropsychopharmacology* **29**, 410–416.

Geldenhuys, W.J., and Van der Schyf, C.J. (2009). The serotonin 5-HT_6 receptor: a viable drug target for treating cognitive deficits in Alzheimer's disease. *Expert Rev. Neurother.* **9**, 1073–1085.

Gerard, C., El Mestikawy, S., Lebrand, C., Adrien, J., Ruat, M., Traiffort, E., Hamon, M., and Martres, M.P. (1996). Quantitative RT-PCR distribution of serotonin 5-HT_6 receptor mRNA in the central nervous system of control or 5,7-dihydroxitriptamine-treated rats. *Synapse* **23**, 164–173.

Gerard, C., Martres, M.P., Lefevre, K., Miquel, M.C., Verge, D., Lanfumey, L., Doucet, E., Hamon, M., and El Mestikawy, S. (1997). Inmuno-localization of serotonin 5-HT_6 receptor-like material in the rat central nervous system. *Brain Res.* **746**, 207–219.

Hannon, J., and Hoyer, D. (2008). Molecular biology of 5-HT receptors. *Behav. Brain Res.* **195**, 198–213.

Hatcher, P.D., Brown, V.J., Tait, D.S., Bate, S., Overend, P., Hagan, J.J., and Jones, D.N. (2005). 5-HT_6 receptor antagonists improve performance in an attentional set shifting task in rats. *Psychopharmacology (Berlin)* **181**, 253–259.

Healy, D.J., and Meador-Woodruff, J.H. (1999). Ionotropic glutamate receptor modulation of 5-HT_6 and 5-HT_7 mRNA expression in rat brain. *Neuropsychopharmacology* **21**, 341–351.

Hirano, K., Piers, T.M., Searle, K.L., Miller, N.D., Rutter, A.R., and Chapman, P.F. (2009). Procognitive 5-HT_6 antagonists in the rat forced swimming test: potential therapeutic utility in mood disorders associated with Alzheimer's disease. *Life Sci.* **84**, 558–562.

Hirst, W.D., Abrahamsen, B., Blaney, F.E., Calver, A.R., Aloj, L., Price, G.W., and Medhurst, A.D. (2003). Differences in the central nervous system distribution and pharmacology of the mouse 5-hydroxytryptamine-$_6$ receptor compared with rat and human receptors investigated by radioligand binding, site-directed mutagenesis, and molecular modeling. *Mol. Pharmacol.* **64**, 1295–1308.

Hirst, W.D., Stean, T.O., Rogers, D.C., Sunter, D., Pugh, P., Moss, S.F., Bromidge, S.M., Riley, G., Smith, D.R., Bartlett, S., Heidbreder, C.A., Atkins, A.R., Lacroix, L.P., Dawson, L.A., Foley, A.G., Regan, C.M., and Upton, N. (2006). SB-399885 is a potent, selective 5-HT_6 receptor antagonist with cognitive enhancing properties in aged rat water maze and novel object recognition models. *Eur. J. Pharmacol.* **553**, 109–119.

Holenz, J., Pauwels, P.J., Díaz, J.L., Mercè, R., Codony, X., and Buschmann, H. (2006). Medicinal chemistry strategies to 5-HT_6 receptor ligands as potential cognitive enhancers and antiobesity agents. *Drug Discov. Today* **11**, 283–299.

Hoyer, D., Hannon, J.P., and Martin, G.R. (2002). Molecular, pharmacological and functional diversity of 5-HT receptors. *Pharmacol. Biochem. Behav.* **7**, 533–554.

Huerta-Rivas, A., González-Espinosa, C., Pérez-García., G, and Meneses, A. (2007). 5-HT_6 receptor mRNA role during memory formation. Society for Neuroscience Meeting, Abstract No. 743.3/AAA26, San Diego, CA, November 3–7.

Huerta-Rivas, A., Perez-Garcia, G., Gonzalez, C., and Meneses, A. (2010). Time-course of 5-HT_6 receptor mRNA expression during memory consolidation and amnesia. *Neurobiol. Learn. Mem.* **93**, 99–110.

Ivachtchenko, A.V., Ivanenkov, Y.A., and Tkachenko, S.E. (2010). 5-hydroxytryptamine subtype 6 receptor modulators: a patent survey. *Expert Opin. Ther. Pat.* **20**, 1171–1196.

Izquierdo, I., Bevilaqua, L.R., Rossato, J.I., Bonini, J.S., Da Silva, W.C., Medina, J.H., and Cammarota, M. (2006). The connection between the hippocampal and the striatal memory systems of the brain: a review of recent findings. *Neurotoxicol. Res.* **10**, 113–121.

Izquierdo, I., Medina, J.H., Viana, M.R.M., Izquierdo, L.A., and Izquierdo, I. (1999). Separate mechanisms of short- and long-term memory. *Behav. Brain Res.* **103**, 1–11.

Jacobs, B.L., and Azmitia, E.C. (1992). Structure and function of the brain serotonin system. *Physiol. Rev.* **72**(1); 165–229.

Johnson, C.N., Ahmed, M., and Miller, N.D. (2008). 5-HT_6 receptor antagonists: prospects for the treatment of cognitive disorders including dementia. *Curr. Opin. Drug. Discov. Dev.* **11**, 642–654.

Kalueff, A.V., Olivier, J.D., Nonkes, L.J., and Homberg, J.R. (2010). Conserved role for the serotonin transporter gene in rat and mouse neurobehavioral endophenotypes. *Neurosci. Biobehav. Rev.* **34**, 373–386.

Kandel, E.R. (2001). The molecular biology of memory storage: a dialogue between genes and synapses. *Science* **294**, 1030–1038.

Kendall, I., Slotten, H.A., Codony, X., Burgueno, J., Pauwels, P.J., Vela, J.M., and Fone, K.C. (2010). E-6801, a 5-HT$_6$ receptor agonist, improves recognition memory by combined modulation of cholinergic and glutamatergic neurotransmission in the rat. *Psychopharmacology (Berlin)* (Epub ahead of print) PMID: 20405281

King, M.V., Marsden, C.A., and Fone, K.C. (2008). A role for the 5-HT$_{1A}$, 5-HT$_4$ and 5-HT$_6$ receptors in learning and memory. *Trends Pharmacol. Sci.* **29**, 482–492.

King, M.V., Sleight, A.J., Woolley, M.L., Topsham, I.A., Marsden, D.A., and Fone, K.C. (2004). 5-HT$_6$ receptors antagonists reverse delay-dependent deficits in novel object discrimination by enhancing consolidation – an effect sensitive to NMDA receptor antagonism. *Neuropharmacology* **47**, 195–204.

King, M.V., Spicer, C.H., Sleight, A.J., Marsden, C.A., and Fone, K.C. (2009). Impact of regional 5-HT depletion on the cognitive enhancing effects of a typical 5-ht$_6$ receptor antagonist, Ro 04-6790, in the Novel Object Discrimination task. *Psychopharmacology (Berlin)* **202**, 111–123.

Koenig, P., Smith, E.E., Troiani, V., Anderson, C., Moore, P., and Grossman, M. (2008). Medial temporal lobe involvement in an implicit memory task: evidence of collaborating implicit and explicit memory systems from FMRI and Alzheimer's disease. *Cereb. Cortex* **18**, 2831–2843.

Li, Z., Huang, M., Prus, A.J., Dai, J., and Meltzer, H.Y. (2007). 5-HT$_6$ receptor antagonist SB-399885 potentiates haloperidol and risperidone-induced dopamine efflux in the medial prefrontal cortex or hippocampus. *Brain Res.* **1134**, 70–78.

Lieben, C.K.J., Blokland, A., Sik, A., Sung, E., van Nieuwenhuizen, P., and Schreiber, R. (2005). The selective 5-HT$_6$ receptor antagonist Ro4368554 restores memory performance in cholinergic and serotonergic models of memory deficiency in the rat. *Neuropsychopharmacology* **30**, 2169–2179.

Lindner, M.D., Hodges Jr., D.B., Hogan, J.B., Orie, A.F., Corsa, J.A., Barten, D.M., Polson, C., Robertson, B.J., Guss, V.L., Gillman, K.W., Starrett Jr., J.E., and Gribkoff, V.K. (2003). An assessment of the effects of serotonin 6 (5-HT$_6$) receptor antagonists in rodent models of learning. *J. Pharmacol. Exp. Ther.* **307**, 682–691.

Linden, R., Martins, V.R., Prado, M.A., Cammarota, M., Izquierdo, I., and Brentani, R.R. (2008). Physiology of the prion protein. *Physiol. Rev.* **88**, 673–728.

Liu, G.L., and Robichaud, A.J. (2009). 5-HT$_6$ antagonists as potential treatment for cognitive dysfunction. *Drug Dev. Res.* **70**, 145–168.

Loiseau, F., Dekeyne, A., and Millan, M.J. (2008). Pro-cognitive effects of 5-HT$_6$ receptor antagonists in the social recognition procedure in rats: implication of the frontal cortex. *Psychopharmacology (Berlin)* **196**, 93–104.

Lorke, D.E., Lu, G., Cho, E., and Yew, D.T. (2006). Serotonin 5-HT$_{2A}$ and 5-HT$_6$ receptors in the prefrontal cortex of Alzheimer and normal aging patients. *BMC Neurosci.* **7**, 36.

Lynch, M.A. (2004). Long-term potentiation and memory. *Physiol. Rev.* **84**, 87–136.

Marcos, B., Cabero, M., Solas, M., Aisa, B., and Ramirez, M.J. (2010). Signaling pathways associated with 5-HT$_6$ receptors: relevance for cognitive effects. *Int. J. Neuropsychopharmacol.* **9**, 1–10.

Marcos, B., García-Alloza, M., Gil-Bea, F.J., Chuang, T.T., Francis, P.T., Chen, C.P., Tsang, S.W., Lai, M.K., and Ramirez, M.J. (2008). Involvement of an altered 5-HT$_6$ receptor function in behavioral symptoms of Alzheimer's disease. *J. Alzheimers Dis.* **14**, 43–50.

Marcos, B., Gil-Bea, F.J., Hirst, W.D., García-Alloza, M., and Ramírez, M.J. (2006). Lack of localization of 5-HT$_6$ receptors on cholinergic neurons: implication of multiple neurotransmitter systems in 5-HT$_6$ receptor-mediated acetylcholine release. *Eur. J. Neurosci.* **24**, 1299–1306.

Martin, C., and Sibson, N.R. (2008). Pharmacological MRI in animal models: a useful tool for 5-HT research? *Neuropharmacology* **55**, 1038–1047.

McGaugh, J.L. (2006). Make mild moments memorable: add a little arousal. *Trends Cogn. Sci.* **10**, 345–347.

Meneses, A. (1999). 5-HT system and cognition. *Neurosci. Biobehav. Rev.* **23**, 1111–1125.

Meneses, A. (2001a). Role of 5-HT_6 receptors in memory formation. *Drug News Perspect.* **14**, 396–400.

Meneses, A. (2001b). Effects of the 5-HT_6 receptor antagonist Ro 04-6790 on learning consolidation. *Behav. Brain Res.* **118**, 107–110.

Meneses, A. (2003). Pharmacological analysis of an associative learning task: 5-HT_1 to 5-HT_7 receptor subtypes function on a Pavlovian/instrumental autoshaped memory. *Learn. Mem.* **10**, 363–372.

Meneses, A. (2007a). Stimulation of 5-HT_{1A}, 5-HT_{1B}, 5-$HT_{2A/2A}$, 5-HT_3 and 5-HT_4 receptors or 5-HT uptake inhibition: short- and long-term memory. *Behav. Brain Res.* **184**, 81–90.

Meneses, A. (2007b). Do serotonin$_{1-7}$ receptors modulate short and long-term memory? *Neurobiol. Learn. Mem.* **87**, 561–572.

Meneses, A., Manuel-Apolinar, L., Castillo, C., and Castillo, E. (2007). Memory consolidation and amnesia modify 5-HT_6 receptors expression in rat brain: an autoradiographic study. *Behav. Brain Res.* **178**, 53–61.

Meneses, A., and Perez-Garcia, G. (2007). 5-HT_{1A} receptors and memory. *Neurosci. Biobehav. Rev.* **3**, 705–727.

Meneses, A., Perez-Garcia, G., Liy-Salmeron, G., Flores-Galvez, D., Castillo, C., and Castillo, E. (2008). The effects of the 5-HT_6 receptor agonist EMD and the 5-HT_7 receptor agonist AS19 on memory formation. *Behav. Brain Res.* **195**, 112–119.

Meneses, A., Perez-Garcia, G., Liy-Salmeron, G., Ponce-Lopez, T., Tellez, R., and Flores-Galvez, D. (2009). Associative learning, memory and serotonin: a neurobiological and pharmacological analysis. In: Rocha Arrieta, L.L., Granados-Soto, V. (Eds.), *Models of Neuropharmacology.* Transworld Research Network, Trivandrum, 978-81-7895-383-0 pp. 169–182.

Meneses, A., Perez-Garcia, G., Ponce-Lopez, T., and Castillo, C. Serotonin transporter and memory. *Neuropharmacology.* 2011 (in press).

Millan, M.J., Marin, P., Bockaert, J., and la Cour, C.M. (2008). Signaling at G-protein-coupled serotonin receptors: recent advances and future research directions. *Trends Pharmacol. Sci.* **29**, 454–464.

Mitchell, E.S., and Neumaier, J.F. (2005). 5-HT_6 receptors: a novel target for cognitive enhancement. *Pharmacol. Ther.* **108**, 320–333.

Mitchell, E.S., McDevitt, R.A., and Neumaier, J.F. (2009). Adaptations in 5-HT receptor expression and function: implications for treatment of cognitive impairment in aging. *J. Neurosci. Res.* **87**, 2803–2811.

Mitchell, E.S., Sexton, T., and Neumaier, J.F. (2007). Increased expression of 5-HT_6 receptors in the rat dorsomedial striatum impairs instrumental learning. *Neuropsychopharmacology* **32**, 1520–1530.

Myhrer, T. (2003). Neurotransmitter systems involved in learning and memory in the rat: a meta-analysis based on studies of four behavioral tasks. *Brain Res. Rev.* **41**, 268–287.

Ögren, S.O. (1985). Evidence for a role of brain serotonergic neurotransmission in avoidance learning. *Acta Physiol Scand Suppl.* **544**, 1–71.

Ögren, S.O., Eriksson, T.M., Elvander-Tottie, E., D'Addario, C., Ekström, J.C., Svenningsson, P., Meister, B., Kehr, J., and Stiedl, O. (2008). The role of 5-HT_{1A} receptors in learning and memory. *Behav. Brain Res.* **195**, 54–77.

Packard, M.G., and Knowlton, B.J. (2002). Learning and memory functions of the basal ganglia. *Ann. Rev. Neurosci.* **25**, 563–593.

Peele, D.B., and Vincent, A. (1989). Strategies for assessing learning and memory, 1978–1987: a comparison of behavioral toxicology, psychopharmacology, and neurobiology. *Neurosci. Biobehav. Rev.* **13**(4); 317–322.

Pérez-García, G., González-Espinosa, C., and Meneses, A. (2006). An mRNA expression analysis of stimulation and blockade of 5-HT_7 receptors during memory consolidation. *Behav. Brain Res.* **169**, 83–92.

Pérez-García, G., and Meneses, A. (2005). Oral administration of the 5-HT$_6$ antagonists SB-357134 and SB-399885 improves memory formation in an autoshaping learning task. *Pharmacol. Biochem. Behav.* **81**, 673–682.

Pérez-García, G., and Meneses, A. (2008a). Memory formation, amnesia, improved memory and reversed amnesia: 5-HT role. *Behav. Brain Res.* **195**, 17–29.

Pérez-García, G., and Meneses, A. (2008b). Ex-vivo study of 5-HT$_{1A}$ and 5-HT$_7$ receptor agonists and antagonists on cAMP accumulation during memory formation and amnesia. *Behav. Brain Res.* **195**, 139–146.

Pérez-García, G., and Meneses, A. (2009). Memory time-course: mRNA 5-HT$_{1A}$ and 5-HT$_7$ receptors. *Behav. Brain Res.* **202**, 102–113.

Radley, J.J., Farb, C.R., He, Y., Janssen, W.G., Rodrigues, S.M., Johnson, L.R., Hof, P.R., LeDoux, J. E., and Morrison, J.H. (2007). Distribution of NMDA and AMPA receptor subunits at thalamo-amygdaloid dendritic spines. *Brain Res.* **1134**, 87–94.

Reid, M., Carlyle, I., Caulfield, W.L., Clarkson, T.R., Cusick, F., Epemolu, O., Gilfillan, R., Goodwin, R., Jaap, D., O'Donnell, E.C., Presland, J., Rankovic, Z., Spinks, D., Spinks, G., Thomson, A.M., Thomson, F., Strain, J., and Wishart, G. (2010). The discovery and SAR of indoline-3-carboxamides-A new series of 5-HT$_6$ antagonists. *Bioorg. Med. Chem. Lett.* **20**(12); 3713–3716.

Riemer, C., Borroni, E., Levet-Trafit, B., Martin, J.R., Poli, S., Porter, R.H.P., and Bös, M. (2003). Influence of the 5-HT$_6$ receptor on acetylcholine release in the cortex: pharmacological characterization of 4-(2-bromo-6-pyrrolidin-1-ylpyridine-4-sulfonyl)phenylamine, a potent and selective 5-HT$_6$ receptor antagonist. *J. Med. Chem.* **46**, 1273–1276.

Roberts, J.C., Reavill, C., East, S.Z., Harrison, P.J., Patel, S., Routledge, C., and Leslie, R.A. (2002). The distribution of 5-HT$_6$ receptors in rat brain: an autoradiographic binding study using the radiolabeled 5-HT$_6$ receptor antagonist [^{125}I]SB-258585. *Brain Res.* **934**, 49–57.

Rossé, G., and Schaffhauser, H. (2010). 5-HT$_6$ receptor antagonists as potential therapeutics for cognitive impairment. *Curr. Top. Med. Chem.* **10**, 207–221.

Roth, B.L., Hanizavareh, S.M., and Blum, A.E. (2004). Serotonin receptors represent highly favorable molecular targets for cognitive enhancement in schizophrenia and other disorders. *Psychopharmacology (Berlin)* **174**, 17–24.

Ruat, M., Traiffort, E., Arrang, J.M., Tardivel-Lacombe, J., Diaz, J., Leurs, R., and Schwartz, J.C. (1993). A novel rat serotonin (5-HT$_6$) receptor: molecular cloning, localization and stimulation of cAMP accumulation. *Biochem. Biophys. Res. Commun.* **193**, 268–276.

Russell, M.G., and Dias, R. (2002). Memories are made of this (perhaps): a review of serotonin 5-HT$_6$ receptor ligands and their biological functions. *Curr. Top. Med. Chem.* **2**, 643–654.

Schechter, L.E., Smith, D.L., Zhang, G.M., Li, P., Lin, Q., Rosenzweig-Lipson, S., Robichaud, A., Bernotas, R., and Beyer, C.E. (2004). WAY-466: pharmacological profile of a novel and selective 5-HT$_6$ agonist. *Int. J. Neuropsychopharmacol.* **7**(Suppl. 1); S291.

Schmitt, J.A., Wingen, M., Ramaekers, J.G., Evers, E.A., and Riedel, W.J. (2006). Serotonin and human cognitive performance. *Curr. Pharm. Des.* **12**, 2473–2486.

Schreiber, R., Vivian, J., Hedley, L., Szczepanski, K., Secchi, R.L., Zuzow, M., van Laarhoven, S., Moreau, J.L., Martin, J.R., Sik, A., and Blokland, A. (2007). Effects of the novel 5-HT$_6$ receptor antagonist RO4368554 in rat models for cognition and sensorimotor gating. *Eur. Neuropsychopharmacol.* **17**, 277–288.

Sleight, A., Boess, F.G., Bös, M., and Bourson, A. (1998). The putative 5-ht$_6$ receptor: localization and function. *Ann. N. Y. Acad. Sci.* **861**, 91–96.

Sossin, W.S. (2008). Defining memories by their distinct molecular traces. *Trends Neurosci.* **31**, 170–175.

Steckler, T., and Sahgal, A. (1995). The role of serotonergic-cholinergic interactions in the mediation of cognitive behaviour. *Behav. Brain Res.* **67**, 165–199.

Stewart, A.J., Fox, A., Morimoto, B.H., and Gozes, I. (2007). Looking for novel ways to treat the hallmarks of Alzheimer's disease. *Expert Opin. Investig. Drugs* **16**, 1183–1196.

Sweatt, J.D. (2009). Experience-dependent epigenetic modifications in the central nervous system. *Biol. Psychiatry* **65**, 191–197.

Tassone, A., Madeo, G., Sciamanna, G., Pisani, A., and Bonsi, P. (2010). Electrophysiology of 5-HT_6 receptors. *Int Rev Neurobiol.* **94**, 111–128.

Tellez, R., Rocha, L., Castillo, C., and Meneses, A. (2010). Autoradiographic study of serotonin transporter during memory formation. *Behav. Brain Res.* **212**(1); 12–26.

Terry Jr., A.V., Buccafusco, J.J., and Wilson, C. (2008). Cognitive dysfunction in neuropsychiatric disorders: selected serotonin receptor subtypes as therapeutic targets. *Behav. Brain Res.* **195**, 30–38.

Tremblay, M.A., Acker, C.M., and Davies, P. (2010). Tau phosphorylated at tyrosine 394 is found in Alzheimer's disease tangles and can be a product of the Abl-related kinase, Arg. *J. Alzheimers Dis.* **19**, 721–733.

Upton, N., Chuang, T.T., Hunter, A.J., and Virley, D.J. (2008). 5-HT_6 receptor antagonists as novel cognitive enhancing agents for Alzheimer's disease. *Neurotherapeutics* **5**, 458–469.

van Praag, H.M. (2004). The cognitive paradox in posttraumatic stress disorder: a hypothesis. *Prog. Neuropsychopharmacol. Biol. Psychiatry* **28**(6); 923–935.

Ward, R.P., Hamblin, M.W., Lachowicz, J.E., Hoffman, B.J., Sibley, D.R., and Dorsa, D.M. (1995). Localization of serotonin subtype 6 receptor messenger RNA in the rat brain by in situ hybridization histochemistry. *Neuroscience* **64**, 1105–1111.

West, P.J., Marcy, V.R., Marino, M.J., and Schaffhauser, H. (2009). Activation of the 5-HT_6 receptor attenuates long-term potentiation and facilitates GABAergic neurotransmission in rat hippocampus. *Neuroscience* **164**, 692–701.

Wilson, C., and Terry, A.V. (2009). Enhancing cognition in neurological disorders: potential usefulness of 5-HT_6 antagonists. *Drugs Future* **34**, 969–975.

Woolley, M.L., Marsden, C.A., and Fone, K.C. (2004). 5-ht_6 receptors. *Curr. Drugs Targets CNS Neurol. Disord.* **3**, 59–79.

Woolley, M.L., Marsden, C.A., Sleight, A.J., and Fone, K.C. (2003). Reversal of a cholinergic-induced deficit in a rodent model of recognition memory by the selective 5-HT_6 receptor antagonist, Ro 04-6790. *Psychopharmacology* **170**, 358–367.

Ye, X., and Carew, T.J. (2010). Small G protein signaling in neuronal plasticity and memory formation: the specific role of ras family proteins. *Neuron* **68**, 340–361.

Yun, H.M., Baik, J.H., Kang, I., Jin, C., and Rhim, H. (2010). Physical interaction of Jab1 with human serotonin 6 G-protein-coupled receptor and their possible roles in cell survival. *J. Biol. Chem.* **285**, 10016–10029.

Yun, H.M., Kim, S., Kim, H.J., Kostenis, E., Kim, J.I., Seong, J.Y., Baik, J.H., and Rhim, H. (2007). The novel cellular mechanism of human 5-HT_6 receptor through an interaction with Fyn. *J. Biol. Chem.* **282**, 5496–5505.

Zola-Morgan, S., and Squire, L.R. (1993). Neuroanatomy of memory. *Ann. Rev. Neurosci.* **168**, 547–563.

BEHAVIORAL PHARMACOLOGY: POTENTIAL ANTIDEPRESSANT AND ANXIOLYTIC PROPERTIES

Anna Wesołowska and Magdalena Jastrzębska-Więsek

Department of Clinical Pharmacy, Jagiellonian University Medical College, 30-688 Kraków, Poland

I. Introduction

The heterogeneous nature of depression (and, to lesser extent, of anxiety) makes successful treatment of the condition even more challenging, and many available medications have shown their limitations in terms of efficacy – as evidenced by a delayed onset of action, unsatisfactory response and remission rates, and residual symptoms (Gaynes *et al.*, 2008; Huynh and McIntyre, 2008; Rush *et al.*, 2006). In addition, there are also safety issues associated with many currently available antidepressant/anxiolytic drugs. Clinical and preclinical considerations have stimulated the investigation and development of numerous combination therapies and novel targets that have been identified, which may demonstrate improvements in the medical treatment of mood disorders, such as depression and anxiety. One of these approaches is directly targeting monoamine receptors, such as the recently discovered serotonin receptor 5-HT_6. Preclinical efforts to evaluate a possible link between the 5-HT_6 receptor and mood disorders have generally been inconclusive yet. The observation that equivalent antidepressant and anxiolytic potency and efficacy can be delivered in animal models by both 5-HT_6 receptor agonists and antagonists does not help in explaining why agonists and/or antagonists of this receptor may serve as potential antidepressant/anxiolytic drugs

INTERNATIONAL REVIEW OF NEUROBIOLOGY, VOL. 96
DOI: 10.1016/B978-0-12-385902-0.00003-6

and which one of them would serve better for treatment of affective disorders. The present commentary focuses on the current state of knowledge regarding antidepressant- and anxiolytic-like activity of 5-HT$_6$ receptor ligands in animal models and their possible mechanism of action. These data, together with supporting and sometimes contradicting evidence, are discussed below.

II. Depression

A. Antidepressant-Like Properties of 5-HT$_6$ Receptor Agonists

The preclinical literature regarding the effects of 5-HT$_6$ receptor agonists in animal models of depression is scarce, mainly due to the paucity of selective compounds. Three of them, that is, WAY-181187 (2-{1-[(6-chloroimidazo[2,1-*b*][1,3]thiazol-5-yl)sulfonyl]-1*H*-indol-3-yl}-*N*,*N*-dimethylethanamine), WAY-208466 (*N*-[2-[3-(3-fluorophenylsulfonyl)-1*H*-pyrrolo[2,3-*b*]pyridin-1-yl]ethyl]-*N*,*N*-dimethylamine) (Schechter *et al.*, 2008), and LY-586713 (de Foubert *et al.*, 2007), have recently been described as novel, selective, and full 5-HT$_6$ receptor agonists. However, the binding studies indicate that WAY-181187 and WAY-208466 show affinity for the 5-HT$_{2C}$ agonist binding site, although WAY-181187 retains approximately 60-fold selectivity for the 5-HT$_6$ receptor (Schechter *et al.*, 2008). As for compound LY-586713, there are no data published demonstrating its affinity for both 5-HT$_6$ receptor and other binding sites as well as confirming its selectivity. Another compound, 11q (N_1-(6-chloroimidazo[2,1-*b*][1,3]thiazole-5-sulfonyl)tryptamine), was shown to belong to a group of high-affinity, potent, and full 5-HT$_6$ receptor agonists (Cole *et al.*, 2007). EMDT (2-ethyl-5-methoxy-*N*,*N*-dimethyltryptamine) (Glennon *et al.*, 2000), ST1936 (*N*,*N*-dimethyl-2-(5-chloro-2-ethyl-1*H*-indol-3-yl)ethylamine) (Borsini *et al.*, 2008), and EMD 386088 (5-chloro-2-methyl-3-(1,2,3,6-tetrahydro-4-pyridinyl)-1*H*-indole) (Mattsson *et al.*, 2005) are successive 5-HT$_6$ receptor agonists used in animal studies to examine the role of 5-HT$_6$ receptor in depression (Borsini *et al.*, 2008; Jastrzębska-Więsek *et al.*, 2010; Svenningsson *et al.*, 2007), but there is some uncertainty about their selectivity.

The selective 5-HT$_6$ receptor agonists WAY-181187 and WAY-208466 were tested for potential antidepressant activity in the modified version of forced swim test in rats (Table I) (Carr *et al.*, 2010). Both agonists, administered three times, produced antidepressant-like effects with the pattern of response similar to the effects of serotonergic compounds (Detke *et al.*, 1995); that is, they significantly lowered immobility and increased swimming behavior without effect on climbing. The antidepressant-like activity of these two compounds were not due to a general increase in locomotor activity, as neither drug produced hyperactivity measured as

Table I
EFFECTS OF 5-HT$_6$ RECEPTOR LIGANDS IN BEHAVIORAL MODELS OF DEPRESSION IN MICE AND RATS.

Compound	Animal Model	Effect Active Doses	Tested Doses	References
Agonists				
WAY-181187	Modified forced swim test in rats	Antidepressant like (3 × 17 mg/kg)	3 × 3–17 mg/kg	Carr *et al.* (2010)
WAY-208466	Modified forced swim test in rats	Antidepressant like (3 × 30 mg/kg)	3 × 7.5–30 mg/kg	Carr *et al.* (2010)
EMDT	Tail suspension test in mice	Antidepressant like (2.5–15 mg/kg)	1–15 mg/kg	Svenningsson *et al.* (2007)
ST1936	Forced swim test in rats	Antidepressant like (100 mg/kg)	10, 30, and 100 mg/kg	Borsini *et al.* (2008)
	Acute escape deficit induced by unavoidable stress exposure	Antidepressant like (repeated dose)		De Montis *et al.* (2009)
	Chronic escape deficit induced by unavoidable stress exposure	No effect of repeated dose		De Montis *et al.* (2009)
	Model of anhedonia	Antianhedonic (repeated dose)		De Montis *et al.* (2009)
EMD 386088	Forced swim test in rats	Antidepressant like (10 mg/kg)	10 mg/kg	Jastrzębska-Więsek *et al.* (2010)
Antagonists				
SB-399885	Tail suspension test in mice	Antidepressant like (10–30 mg/kg)	3–30 mg/kg	Wesołowska and Nikiforuk (2007)
	Forced swim test in mice	Antidepressant like (20 and 30 mg/kg)	10–30 mg/kg	Wesołowska and Nikiforuk (2007)
	Forced swim test in rats	Antidepressant like (1 × 10 mg/kg)	3–20 mg/kg	Wesołowska (2007), Wesołowska and Nikiforuk (2007)
		Antidepressant like (3 × 3 or 10 mg/kg)	3 × 0.3–10 mg/kg	Hirano *et al.* (2009)
SB-399885 + antidepressants[a]	Forced swim test in rats	Antidepressant like	3 mg/kg + antidepressant	Wesołowska and Nikiforuk (2008)

(*continued*)

Table I (*continued*)

Compound	Animal Model	Effect Active Doses	Tested Doses	References
SB-399885 + citalopram[b]	Forced swim test in rats	No effect	3 mg/kg + 20 mg/kg	Wesołowska and Nikiforuk (2008)
SB-271046	Tail suspension test in mice	No effect	1–10 mg/kg	Svenningsson *et al.* (2007)
	Forced swim test in rats	Antidepressant like (3 × 10 or 30 mg/kg)	3 × 3–30 mg/kg	Hirano *et al.* (2009)
SB-258585[c]	Forced swim test in rats	Antidepressant like (3 μg)	1–10 μg	Wesołowska *et al.* (2007)

[a] Synergistic enhancement when using a combination of inactive doses of SB-399885 together with imipramine, desipramine, bupropion, or moclobemide.
[b] Combination of nonactive doses of SB-399885 and citalopram.
[c] SB-258585 administered into CA1 region of hippocampus.

distance traveled in the open field (Carr *et al.*, 2010). However, the latter authors did not try to modulate the antidepressant-like effects of both compounds by using a selective 5-HT_6 receptor antagonist in order to confirm that their activity observed in the forced swim test really is mediated by the 5-HT_6 receptor.

Svenningsson *et al.* (2007) showed the ability of a 5-HT_6 receptor agonist EMDT, administered acutely, to mimic some of antidepressant-like biochemical and behavioral effects of fluoxetine. EMDT, like acute fluoxetine, increased the phosphorylation state of Thr^{34}-DARPP-32 both in brain slices and in the intact brain, phospho-Ser^{845}-GluR1, as well as the expression of c-*fos* mRNA throughout the striatum and cerebral cortex. Moreover, in a similar range of doses, EMDT significantly reduced the immobility time of mice in the tail suspension test (Table I). A selective 5-HT_6 receptor antagonist SB-271046 (5-chloro-*N*-[4-methoxy-3-(1-piperazinyl)phenyl]-3-methylbenzothiophene-2-sulfonamide), administered at a dose that by itself produced no effect in those assays, completely blocked the biochemical and behavioral antidepressant-like effects produced by EMDT and partially blocked all these effects produced by fluoxetine (Svenningsson *et al.*, 2007). Because fluoxetine has only low-to-moderate affinity for 5-HT_6 receptors (Monsma *et al.*, 1993), the authors concluded that the inhibitory action of SB-271046 on fluoxetine-mediated effects involves blocking 5-HT_6 receptor activation that had been elicited by the fluoxetine-induced elevations of extracellular levels of serotonin and does not involve the direct competition of either compound at 5-HT_6 sites (Svenningsson *et al.*, 2007). In the case of compound EMDT that has nanomolar affinity for the 5-HT$_6$ receptor (K_i = 16 nM, Glennon *et al.*, 2000), its antidepressant-like activity seems to be connected with stimulation of this receptor.

The recent congress abstract has introduced another 5-HT$_6$ receptor agonist, compound ST1936 that possesses nanomolar affinity for 5-HT$_6$ receptors (K_i = 48 nM) and produced an antidepressant-like effect in the forced swim test in rats (Borsini *et al.*, 2008) (Table I); however, its activity in that model has not been attenuated by a selective 5-HT$_6$ receptor antagonist SB-271046 (Borsini, personal information), which can indicate that, in the case of this compound, the potential antidepressant activity observed in the forced swim test results from different mechanism than stimulation of the 5-HT$_6$ receptor. Nevertheless, repeated administration of ST1936 prevented the development of escape deficit induced by acute unavoidable stress exposure, but unlike classical antidepressants, did not revert the escape deficit induced by a chronic stress exposure. Additionally, ST1936, when administered chronically, effectively counteracted chronic stress-induced anhedonia and restored the capacity of stressed rats to acquire vanilla sugar (De Montis *et al.*, 2009) (Table I). There are no data showing whether this antianhedonic activity of ST1936 is modulated by the 5-HT$_6$ receptor yet, although experiments with this compound are ongoing.

Mattsson *et al.* (2005) have presented EMD 386088 that displayed high affinity for 5-HT$_6$ receptors (IC_{50} = 7.4) and a moderate one for 5-HT$_3$ ones (IC_{50} = 34). Functionally, when tested by cAMP accumulation assay, EMD 386088 showed features of a 5-HT$_6$ receptor agonist. In our laboratory, this compound exhibited antidepressant-like activity in the forced swim test in rats when given acutely (Table I) (Jastrzębska-Więsek *et al.*, 2010). On this stage of experimentation, one cannot say about mechanism of its action because experiments are in progress.

As has been mentioned above, the selective agonists produce potential antidepressant activity and a selective 5-HT$_6$ receptor antagonist SB-271046 blocked an antidepressant-like effect of EMDT only (Svenningsson *et al.*, 2007). It is well established that the shortening of immobility time, induced by antidepressant drugs and observed in behavioral tests of despair, depends on the enhancement of the central serotonin and catecholamine neurotransmission (Borsini, 1995; Cryan *et al.*, 2005; Porsolt *et al.*, 1979). Unfortunately, no information is available on the effect of EMDT on the levels of serotonin and catecholamines. A microdialysis study has only shown that a selective 5-HT$_6$ receptor agonist, WAY-181187, decreased the basal release of serotonin, dopamine, and norepinephrine in the frontal cortex, striatum, and amygdala of freely moving rats (Schechter *et al.*, 2008). The latter effects were found to be blocked by local infusion of a $GABA_A$ receptor antagonist, bicuculline, which confirms a relationship between the 5-HT$_6$ receptor and GABAergic system and is entirely consistent with a study showing dense colocalization of the 5-HT$_6$ receptor and the GABA synthesizing enzyme, glutamic acid decarboxylase, in the rat cortex, hippocampus, and striatum (Ward and Dorsa, 1996; Woolley *et al.*, 2004). Additionally, acute administration of

WAY-181187 increased the extracellular levels of GABA in the rat frontal cortex, dorsal hippocampus, striatum, and amygdala, without altering basal levels of glutamate. However, in hippocampal slices, pretreatment with WAY-181187 significantly and dose-dependently attenuated sodium azide-stimulated increases in glutamate concentrations (Schechter *et al.*, 2008). Impairment of GABAergic neurotransmission has been described in patients with a variety of depressive illness (Leung and Xue, 2003; Pierce *et al.*, 1996; Sanacora *et al.*, 2000). Moreover, Sanacora *et al.* (2004) observed higher glutamate levels in depressed patients compared with healthy age-matched controls. The preclinical studies and preliminary clinical trials with ketamine, an NMDA receptor antagonist (Paul and Skolnick, 2003; Yilmaz *et al.*, 2002; Zarate *et al.*, 2006), demonstrated its potential antidepressant activity, which supports the notion that attenuating glutamatergic neurotransmission can be beneficial. Thus, improvement of GABAergic neurotransmission in connection with dampening-stimulated glutamatergic transmission could be proposed as a mechanism of the antidepressant-like activity of selective 5-HT_6 receptor agonists, at least WAY-181187.

In view of lack of selectivity of compounds ST1936 and EMD 386088 for 5-HT_6 receptors and lack of more behavioral results, on the current stage of knowledge one cannot consider their mechanism of action detected in the forced swim test. Although the microdialysis results presented by Di Chiara *et al.* (2009) indicate that ST1936 dose-dependently increased the extracellular level of dopamine and norepinephrine in the rat prefrontal cortex and the nucleus accumbens shell and core, without affecting the serotonin level. Moreover, its above-mentioned effects were prevented by pretreatment with SB-271046, a selective 5-HT_6 receptor antagonist (Di Chiara *et al.*, 2009).

Several independent studies have demonstrated that repeated, but not acute, administration of antidepressants increases the expression of genes whose corresponding proteins regulate synaptic plasticity in the brain (Coppell *et al.*, 2003; de Foubert *et al.*, 2004; Nestler *et al.*, 2002; Russo-Neustadt and Chen, 2005). As a result, a brain-derived neurotrophic factor (BDNF) and the effector immediate early gene activity-regulated cytoskeletal associated protein (*Arc*) have been identified as possible targets for antidepressant action (Pei *et al.*, 2003; Shimizu *et al.*, 2003; Shirayama *et al.*, 2002). Experiments conducted by de Foubert *et al.* (2007) provided the first evidence for the involvement of the 5-HT_6 receptors in regulating BDNF and *Arc* mRNA expression. They showed that a selective 5-HT_6 receptor agonist, compound LY-586713, when administered acutely, caused marked increases of BDNF and *Arc* mRNA levels in hippocampal and cortical regions. These increases were attenuated by SB-271046 in all regions of the hippocampus and the parietal cortex. In conclusion, the above-mentioned results may suggest that direct 5-HT_6 receptor activation results in a more rapid rise in BDNF and *Arc* mRNA expression, which does not require repeated administration, as is the case for some antidepressants (Coppell *et al.*, 2003; de Foubert *et al.*,

2004; Nestler *et al.*, 2002; Russo-Neustadt and Chen, 2005). Because LY-586713 is described as a selective 5-HT$_6$ receptor agonist, it is plausible that the 5-HT$_6$ receptor-mediated onset of BDNF gene expression following a single dose of LY-586713 is directly linked to the activation of 5-HT$_6$ receptors. Subsequently, increases in cAMP levels lead to activation of CREB, a known transcription factor for the BDNF gene (de Foubert *et al.*, 2007). On the other hand, the same authors showed that a selective 5-HT$_6$ receptor antagonist, SB-271046 *per se*, produced increases in *Arc* mRNA levels of magnitude similar to those of LY-586713 in both cingulate and orbital cortex (de Foubert *et al.*, 2007). The above-mentioned results confirm that both 5-HT$_6$ receptor agonist and antagonist may produce similar effects.

B. Antidepressant-Like Properties of 5-HT$_6$ Receptor Antagonists

The first hints about potential antidepressant activity of 5-HT$_6$ receptor antagonists might date back papers written by Kohen *et al.* (2001), Monsma *et al.* (1993), and Sebben *et al.* (1994), which showed that several tricyclic antidepressants such as amitryptiline, and atypical antidepressants such as mianserin, display high affinity and antagonistic activity at 5-HT$_6$ receptors. These data together with the localization of 5-HT$_6$ receptors in limbic and cortical brain areas (Boess *et al.*, 1997, 1998) have led to the implication of a role for 5-HT$_6$ receptor antagonists in the treatment of affective disorders. Moreover, preliminary genetic studies revealed that bipolar affective disorder might be associated with variation in a 5-HT$_6$ gene (Vogt *et al.*, 2000).

These results have encouraged scientists to conduct experiments with selective 5-HT$_6$ receptor antagonists when they have become available. Among them, SB-271046, SB-399885 (*N*-(3,5-dichloro-2-methoxyphenyl)-4-methoxy-3-(1-piperazinyl)benzenesulfonamide), and SB-258585 (4-iodo-*N*-[4-methoxy-3-(4-methyl-1-piperazinyl)phenyl]benzenesulfonamide) are the most widely used, selective, and potent compounds (Bromidge *et al.*, 1999; Hirst *et al.*, 1999, 2000, 2006). It has been shown that SB-399885 exerted an antidepressant-like effect in the forced swim and tail suspension tests in mice. These effects seem to be specific because SB-399885, when administered at antidepressant doses, did not stimulate locomotor activity in mice (Wesołowska and Nikiforuk, 2007) (Table I). In the forced swim test in mice, an anti-immobility effect of SB-399885 was dose dependent and comparable to that of imipramine, which was used as a reference antidepressant drug in that study, whereas in the tail suspension test the potential antidepressant activity of the 5-HT$_6$ receptor antagonist used was weaker than the effect observed after treatment with imipramine and was not dose dependent (Wesołowska and Nikiforuk, 2007).

It is well established that acute administration of antidepressant drugs reduces immobility in behavioral tests of despair; that is, the forced swim and the tail suspension tests, and increases expression of the immediate early gene c-*fos* mRNA in the brain (Beck, 1995; Cryan *et al.*, 2005; Horowitz *et al.*, 2003). Compound SB-271046, another 5-HT_6 receptor antagonist, did not change immobility behavior of mice in the tail suspension test as fluoxetine did. In agreement with behavioral data, treatment with SB-271046 alone had no effect on c-*fos* mRNA expression in certain limbic regions of the frontal cerebral cortex, including the cingulated cortex and the endopiriform cortex (Svenningsson *et al.*, 2007) (Table I). The reason for the discrepancies in the action of two selective 5-HT_6 receptor antagonists and chemical analogs in the mouse tail suspension test is unclear. It should be noted that both experiments were performed on C57BL/6J mice. Although the breeding institutions providing male mice were different (Svenningsson *et al.*, 2007; Wesołowska and Nikiforuk, 2007), the behavioral outcome should have been similar, even though mice are not an ideal species in which to conduct behavioral or physiological experiments pertaining to 5-HT_6 receptor functions. As described by Hirst *et al.* (2003a), the mouse is unique compared with the rat, pig, and man, all of which express relatively high levels of 5-HT_6 receptors. Hence, Svenningsson *et al.* (2007) confirmed in their study that SB-271046 has a high affinity for 5-HT_6 receptors in the mouse forebrain by showing that it displaced specific [^{125}I]-SB-258585 binding at nanomolar concentrations (EC_{50} = 4 nM). Nevertheless, it is noteworthy that doses used for both compounds and experimental protocols were inconsistent. SB-399885 produced a pretty weak, but similar, anti-immobility effect at all doses used, that is, 10, 20, and 30 mg/kg (Wesołowska and Nikiforuk, 2007), whereas SB-271046 was given up to a dose of 10 mg/kg only; higher doses were not tested (Svenningsson *et al.*, 2007). Hirano *et al.* (2009) and Hirst *et al.* (2003b) demonstrated in *ex vivo* binding assays that administration of SB-399885 significantly inhibited specific [^{125}I]-SB-258585 binding in the rat striatum with doses lower than those required for SB-271046 to significantly occupy brain 5-HT_6 receptors. The above results prove the higher efficacy of SB-399885 in comparison to that of SB-271046, at least in rats. The improved potency of SB-399885 seems to be due to an increase in brain penetration when compared to SB-271046, where it shows a threefold improvement in the brain: plasma ratio of 5–15% (Hirst *et al.*, 2006; Routledge *et al.*, 2000). Moreover, the duration of the tail suspension tests also varied: 6 min for SB-399885 and 5 min for SB-271046.

In the forced swim test in rats, the findings for both 5-HT_6 receptor antagonists are consistent. Treatment with an acute dose of 10 mg/kg of SB-399885 produced specific antidepressant-like activity by shortening the immobility time without a stimulatory effect on exploratory activity observed in the open field (Wesołowska, 2007; Wesołowska and Nikiforuk, 2007). The absence of a potential antidepressant action after administration of a higher dose of SB-399885, that is, 20 mg/kg, may be due to a sedative effect, as was evident in the open field test in rats (Wesołowska and Nikiforuk, 2007). In line with that study, Hirano *et al.* (2009) demonstrated antidepressant-like activity for both SB-399885 and SB-271046 administered three times in rats (Table I). In agreement with brain 5-HT_6 receptor occupancy (62 and 96% for 3 and 10 mg/kg of SB-399885, respectively; 56 and 84% for 10 and 30 mg/kg of SB-271046, respectively) (Hirano *et al.*, 2009) and better brain penetration (15% for SB-399885 vs. 5% for SB-271046) (Hirst *et al.*, 2006), SB-399885 was approximately three times more potent than SB-271046 in the forced swim test in rats. Both SB-399885 and SB-271046 significantly suppressed immobility behavior of rats with doses of 3 and 10 mg/kg, respectively; their effects were slightly weaker than that of imipramine (Hirano *et al.*, 2009).

The antidepressant-like activity of SB-399885 in mice and rats and such activity of SB-271046 in rats are most probably connected with their 5-HT_6 receptor antagonistic properties, since both compounds are selective ligands and blockers of 5-HT_6 sites (Bromidge *et al.*, 1999; Hirst *et al.*, 2006; Routledge *et al.*, 2000). Hence, direct involvement of other receptors in their effect ought to be excluded.

The shortening of immobility time, induced by antidepressants in behavioral tests of despair, depends on the enhancement of the central serotonin and catecholamine neurotransmission (Borsini, 1995; Borsini and Meli, 1990; Cryan *et al.*, 2005; Porsolt *et al.*, 1979). Unfortunately, no information is available on the effect of SB-399885 on the basal levels of serotonin. Yet, a microdialysis study has shown that SB-271046 had no influence on the basal release of serotonin in the rat prefrontal cortex, frontal cortex, striatum, and hippocampus (Dawson *et al.*, 2000, 2001; Lacroix *et al.*, 2004; Loiseau *et al.*, 2008). Moreover, 5-HT_6 receptor antagonists, SB-399885 and/or SB-271046, significantly increase cortical and hippocampal extracellular concentrations of dopamine, norepinephrine, and acetylcholine in freely moving rats (Hirst *et al.*, 2003b, 2006; Lacroix *et al.*, 2004; Li *et al.*, 2007; Loiseau *et al.*, 2008). They can also potentiate the behavioral effects of amphetamine (Dawson *et al.*, 2003; Frantz *et al.*, 2002; Pullagurla *et al.*, 2004). Furthermore, high levels of 5-HT_6 receptor immunoreactivity have been found in dopaminergic areas (Gerard *et al.*, 1997). The microdialysis findings have been extended by a study of Wesołowska *et al.* (2007), which demonstrated that administration of *p*-chloramphetamine under their laboratory conditions reduced cortical and hippocampal concentrations of serotonin and its metabolite, and did not modify the antidepressant-like effect of SB-399885 in the forced swim test in rats. However, its anti-immobility action in that test was abolished by the preferential D_1- and D_2-like receptor antagonists SCH-23390 and sulpiride, respectively, and by the α_2-adrenoceptor antagonist idazoxan, but not by prazosin, a blocker of α_1-adrenoceptors (Wesołowska, 2007). On the basis of microdialysis and behavioral studies, it could be proposed that the anti-immobility effect of 5-HT_6 receptor antagonists does not actually require any integrity of serotonin neurons and seems to be connected with activation of dopamine and norepinephrine systems – via D_1- and D_2-like receptors and α_2-adrenoceptors.

It is well documented that 5-HT$_6$ receptor antagonists, including SB-399885, increase extracellular levels of acetylcholine in the hippocampus and prefrontal cortex in freely moving rats (Hirst *et al.*, 2006; Riemer *et al.*, 2003). Acetylcholine is of great importance in cognitive function and diseases, including Alzheimer's disease (Sarter and Parikh, 2005); however, its influence on emotional regulation is less known. Janowsky *et al.* (1972) and Shytle *et al.* (2002) postulated that there is relative or actual cholinergic hyperactivity in depression. Moreover, in animal models cholinolytic drugs enhance the potential antidepressant activity of imipramine (Popik *et al.*, 2003) and suppress immobility behavior of mice in the forced swim and the tail suspension tests (Borsini and Meli, 1988; Cryan *et al.*, 2005). Thus, these findings indirectly suggest that it is unlikely that the antidepressant-like effect of 5-HT$_6$ receptor antagonists observed in animal models develops as a consequence of enhanced acetylcholine release.

The behavioral evidence indicates that the selective blockade of 5-HT$_6$ receptors evoked by SB-399885 may facilitate the anti-immobility effects of antidepressant drugs whose mechanism of action is connected with the inhibition of norepinephrine/dopamine uptake as well as with the inhibition of monoamine oxidase-A (Wesołowska and Nikiforuk, 2008) (Table I). Combining a subefficacious dose of SB-399885 with ineffective doses of imipramine, desipramine, bupropion, or moclobemide results in a pronounced decrease in immobility time in the forced swim test in rats. Only citalopram injected in a nonactive dose jointly with SB-399885 did not induce any effects characteristic of antidepressants in that test. However, no data are available so far on the effect of SB-399885 on antidepressants' blood–brain barrier penetration and pharmacokinetics parameters; thus, pharmacokinetic interaction between a selective 5-HT$_6$ antagonist and antidepressants studied cannot be ruled out. All in all, the above-cited behavioral and neurochemical results seem to point to an involvement of enhanced monoaminergic, namely dopaminergic and/or noradrenergic, neurotransmission in the antidepressant-like activity of a 5-HT$_6$ antagonist (Wesołowska and Nikiforuk, 2008).

In recent years, clinical and laboratory experiments have strengthened the role of the hippocampus in the pathophysiology of depression (Hensler, 2006). Multiple imaging studies have revealed hippocampal volume reductions in depressed patients, which are changes that correlate well with disease duration and executive dysfunction (Lange and Irle, 2004; Neumeister *et al.*, 2005; O'Brien *et al.*, 2004; Saylam *et al.*, 2006; Videbech and Ravnkilde, 2004; Winter and Irle, 2004). Furthermore, it seems that the hippocampus may play an important role in the action of antidepressants as well. It has been demonstrated that treatment of patients with major depressive disorder using these drugs significantly improved memory and depressive symptoms without altering hippocampal volume, which could suggest that antidepressants improve hippocampal function in the absence of detectable structural changes (Vythilingam *et al.*, 2004). An involvement of the hippocampus in the action of antidepressant drugs has also been demonstrated in animal studies. For example, imipramine (Przegaliński *et al.*, 1997) or desipramine (Kostowski, 1985) injected into the rat hippocampus was able to reduce immobility in the forced swim test. Moreover, electrolytic lesion of that structure completely abolished the effect of peripherally administered desipramine (Kostowski, 1985). In addition, Sherman and Allers (1980) found a strong correlation between imipramine concentration in the rat hippocampus and the positive effect of the drug in the behavioral learned helplessness test, an animal model predictive of antidepressant potential. Finally, the finding that antidepressants increase hippocampal BDNF expression seems to support the idea that hippocampal neurogenesis is involved in antidepressant action (Malberg and Schechter, 2005).

The 5-HT$_6$ receptors are present in the hippocampus, including its CA1 region (Gerard *et al.*, 1996, 1997; Monsma *et al.*, 1993; Ward *et al.*, 1995; Yoshioka *et al.*, 1998). Furthermore, 5-HT$_6$ receptor expression appears to be regulated by glucocorticoids. Both adrenalectomy and blockade of glucocorticoid synthesis by metyrapone or aminoglutethimidine treatment selectively upregulated the 5-HT$_6$ receptor mRNA in the CA1 and CA3 pyramidal cells of the rat hippocampus (Yau *et al.*, 1997). As metyrapone and aminoglutethimidine treatments have been used in the clinic to treat resistant depression (Kramlinger *et al.*, 1985; Murphy *et al.*, 1991), the authors speculated that this effect might involve 5-HT$_6$ receptors. Additionally, the hippocampus has been proposed as one of the neuroanatomical structures involved in antidepressant-like activity of a selective 5-HT$_6$ receptor antagonist, SB-258585 (Wesołowska *et al.*, 2007). In fact, recent behavioral results demonstrated that SB-258585, when injected into the CA1 region of the rat hippocampus, produced an antidepressant-like effect in the forced swim test (Table I). Its potential antidepressant activity cannot be attributed to changes in general activity as this drug, given at a dose producing an anti-immobility effect, did not change exploratory locomotor activity measured in the open field test in rats. A noteworthy fact is that the antidepressant-like effect of SB-258585 was comparable with that of imipramine (Wesołowska *et al.*, 2007).

III. Anxiety

A. Anxiolytic-Like Properties of 5-HT$_6$ Receptor Agonists

It has been suggested by Schechter *et al.* (2008) that selective 5-HT$_6$ receptor agonists may play a potential therapeutic role in the treatment of some types of anxiety-related disorders. Thus, WAY-181187 and WAY-208466 (Schechter *et al.*, 2008), as well as compound 11q (Cole *et al.*, 2007), when administered acutely, effectively decreased water intake by rats that had not been water deprived in the

Table II
EFFECTS OF 5-HT$_6$ RECEPTOR LIGANDS IN ANIMAL BEHAVIORAL MODELS OF ANXIETY IN MICE AND RATS.

Compound	Animal Model	Effect Active Doses	Tested Doses	References
Agonists				
WAY-181187	Schedule-induced polydipsia test in rats	Anxiolytic like (178 mg/kg)	56–178 mg/kg	Schechter *et al.* (2008)
	Defensive burying test in rats	Anxiolytic like (10 and 17 mg/kg)	10 and 17 mg/kg	Carr *et al.* (2010)
	Novelty-induced hypophagia in rats	No effect	10 and 17 mg/kg	Carr *et al.* (2010)
WAY-208466	Defensive burying test in rats	Anxiolytic like (15 mg/kg)	15 and 30 mg/kg	Carr *et al.* (2010)
	Novelty-induced hypophagia in rats	Anxiolytic like (15 mg/kg)	15 and 30 mg/kg	Carr *et al.* (2010)
11q	Schedule-induced polydipsia test in rats	Anxiolytic like (178 mg/kg)	56–178 mg/kg	Cole *et al.* (2007)
Antagonists				
SB-399885	Four-plate test in mice	Anxiolytic like (3–20 mg/kg)	1–20 mg/kg	Wesołowska and Nikiforuk (2007)
	Elevated plus maze in rats	Anxiolytic like (0.3–3 mg/kg)	0.1–3 mg/kg	Wesołowska and Nikiforuk (2007)
	Conflict drinking test in rats	Anxiolytic like (1 and 3 mg/kg)	0.3–3 mg/kg	Wesołowska (2008), Wesołowska and Nikiforuk (2007)
SB-399885 + diazepam[a]	Conflict drinking test in rats	Anxiolytic like	0.3 mg/kg + 2.5 mg/kg	Wesołowska (2008)
SB-258585[b]	Conflict drinking test in rats	Anxiolytic like (1 μg)	0.3–3 μg	Wesołowska *et al.* (2007)

[a] Synergistic enhancement when using a combination of inactive doses of SB-399885 together with diazepam.

[b] SB-258585 administered into CA1 region of hippocampus.

schedule-induced polydipsia test (Table II), a model considered to be predictive for efficacy in obsessive compulsive disorder (Hogg and Dalvi, 2004). What is more, pharmacological studies have demonstrated that selective serotonin reuptake inhibitors can also decrease adjunctive drinking in this test (Woods *et al.*, 1993); however, they have to be administered chronically to correspond with their clinical efficacy. Furthermore, WAY-208466 and WAY-181187 reduced burying behavior of rats in the defensive burying test and WAY-208466 decreased approach latency in the novelty-induced hypophagia test (Carr *et al.*, 2010) (Table II). Their anxiolytic-like effects observed in the defensive burying test were similar to the effects produced by established anxiolytic drugs (Bodnoff *et al.*, 1989; Treit *et al.*, 1981) as well as by chronic treatment with antidepressants (Bechtholt *et al.*, 2008; Bodnoff *et al.*, 1988; Bondi *et al.*, 2007; Dulawa *et al.*, 2004). Despite slight differences in the behavioral responses produced by both 5-HT_6 receptor agonists in the novelty-induced hypophagia test (Carr *et al.*, 2010) that may result, as authors suggested, from their different affinity for the 5-HT_{2C} receptors, anxiolytic-like effects, especially observed in the defensive burying test, seem to be specific since WAY-208466 and WAY-181187 did not evoke any significant alterations in general locomotor activity, reactivity to shock, or latency to contact probe (Carr *et al.*, 2010).

In addition to behavioral data, neurochemical results revealed that selective 5-HT_6 receptor agonists increased extracellular levels of GABA in several areas of the rat brain associated with affective disorders (Schechter *et al.*, 2008). Thus, an acute administration of WAY-181187 increased extracellular GABA concentrations in the rat frontal cortex, dorsal hippocampus, striatum, and amygdala; all effects were blocked by a selective 5-HT_6 receptor antagonist SB-271046. In addition to the acute effects, a 14-day treatment with WAY-208466 resulted in robust elevations in the extracellular levels of GABA in the rat dorsolateral frontal cortex, an effect similar in terms of magnitude and duration to that produced by WAY-181187 in the same brain region. These findings highlight the fact that chronic activation of 5-HT_6 receptors does not evoke desensitization. Moreover, employing *in vivo* microdialysis techniques, Schechter *et al.* (2008) have also shown that WAY-181187 did not alter basal glutamate levels, but this 5-HT_6 receptor agonist attenuated sodium azide-stimulated glutamate release in hippocampal slices.

B. Anxiolytic-Like Properties of 5-HT_6 Receptor Antagonists

Few studies using 5-HT_6 receptor antisense oligonucleotides have explored the involvement of 5-HT_6 receptors in rodent models of anxiety, and the reported results have been inconsistent. In the Yoshioka *et al.* (1998) study, 7 days of 5-HT_6 receptor-directed antisense oligonucleotides treatment caused a 30% reduction in [^{3}H]-LSD binding sites, accompanied by a reduction in conditioned fear stress-induced serotonin release in the rat prefrontal cortex, which is suggestive of an anxiolytic-like response. Conversely, Hamon *et al.* (1999) and Otano *et al.* (1999) demonstrated anxiogenic-like activity following chronic administration of the same oligonucleotide sequence in two alternative rat models of anxiety; that is, the elevated plus maze and the social interaction tests, respectively. 5-HT$_6$ receptor-knockout mice, however, displayed few phenotypic abnormalities, except for differences in open field anxiety-related behaviors and in an elevated zero maze; that is, there were no differences in total open quadrant dwell time, number of transitions between open and sheltered maze quadrants, or head dips (Bonasera *et al.*, 2006). A major concern with data arising from this approach is that different developmental compensations may mask the true function of the receptor deleted, as is observed with other serotonin receptor knockouts (Ase *et al.*, 2000). Additionally, the very low expression of 5-HT$_6$ receptors in the mouse brain (Hirst *et al.*, 2003a) should call into question the value of using this species for basic examination of 5-HT$_6$ receptor function.

The development of selective 5-HT$_6$ receptor antagonists has provided useful tools for pharmacological studies that have shed light on the importance of 5-HT$_6$ receptor blockade in anxiety assays. Recently, Wesołowska and Nikiforuk (2007) have observed that a potent and selective 5-HT$_6$ receptor antagonist SB-399885 produced specific anxiolytic-like activity in animal models of anxiety (Table II). It dose-dependently and significantly increased the number of shocks accepted in the conflict drinking (Vogel) test in rats, a model that is widely used and considered to be one of the most specific methods for the detection of potential anxiolytic activity (Millan, 2003), without an effect on either the shock threshold or unpunished water consumption. This finding is supported by results obtained in an elevated plus maze test in rats, a procedure based on rodents' natural aversion to heights and open space. In this model, SB-399885 dose-dependently increased the percentage of time spent in and the number of entries into the open arms of the maze, without stimulating the general exploratory activity of rats detected in the open field test. Both effects observed indicate that there has been a decrease in anxiety after administration of SB-399885. Moreover, SB-399885 also had antianxiety-like activity in the four-plate test in mice (Wesołowska and Nikiforuk, 2007). It is noteworthy that, quantitatively, the potential anxiolytic effect of a 5-HT$_6$ receptor antagonist tested in all three models employed was approximately equivalent to diazepam, particularly in rats.

The potential anxiolytic-like effect of SB-399885 was not modified in rats whose serotonin neurons were destroyed by prior administration of *p*-chloramphetamine (Wesołowska, 2008), pointing at the possibility that the above-mentioned activity is not conditioned by the integrity of serotonin neurons. Such a concept is in line with neuroanatomical results that show that 5-HT$_6$ receptors are located outside serotonin neurons (Gerard *et al.*, 1997; Ward *et al.*, 1995). Thus, the anticonflict activity of SB-399885 observed in the Vogel test in rats does not depend on serotoninergic neurotransmission. The study described by Wesołowska (2008) has also demonstrated that a selective 5-HT$_6$ receptor antagonist tested helped to reveal the anticonflict action of diazepam, either being used in a nonactive dose (Table II). The additive effect of SB-399885 and diazepam may be regarded as a result of pharmacodynamic and/or pharmacokinetic interaction. Because the levels of SB-399885 and diazepam administered alone or jointly have not been analyzed, one cannot rule out a pharmacokinetic interaction at this stage of experimentation. Furthermore, the anticonflict activity of SB-399885 was reduced by the benzodiazepine receptor antagonist flumazenil, which was used at a dose reported to antagonize diverse effects of diazepam, including its anxiolytic-like effect (Boast *et al.*, 1983; File *et al.*, 1985; Liljequist and Engel, 1984; Wesołowska, 2008). Since SB-399885 exhibits no affinity for GABA and benzodiazepine receptors (Hirst *et al.*, 2006), its antianxiety-like effect stems from a functional interaction between 5-HT$_6$ receptors and the GABA/benzodiazepine system. Such a conclusion is supported by neuroanatomical findings concerning the expression of 5-HT$_6$ receptor mRNA on GABAergic neurons (Gerard *et al.*, 1996; Ward and Dorsa, 1996). Moreover, Benes *et al.* (2004) demonstrated an increase in 5-HT$_6$ receptor mRNA in rat hippocampus after GABAergic transmission had been interrupted by a local infusion of picrotoxin. It has also been shown that another 5-HT$_6$ receptor antagonist, SB-357134, induced a concentration-dependent increase in the K^+-evoked GABA efflux in rat striatal slices (Marcos *et al.*, 2006).

Unfortunately, there is no information about the action of SB-399885 on the release of GABA in animal brain areas. However, it has been established that 5-HT$_6$ receptor antagonists, including SB-399885, increase the activity of the acetylcholine system (Foley *et al.*, 2004; Hirst *et al.*, 2006; Jayarajan *et al.*, 2005; Lieben *et al.*, 2005; Riemer *et al.*, 2003; Shirazi-Southall *et al.*, 2002). As has been presented in an excellent review written by Millan (2003), cholinergic pathways do not play a pivotal role in the control of anxiety; however, some experimental evidence indicates that stimulation of cholinergic transmission is accompanied by anxiolytic-like activity in animals (Balerio *et al.*, 2006; Brioni *et al.*, 1993; Degroot and Treit, 2002). Thus, involvement of the cholinergic system in the anxiolytic-like activity of SB-399885 and in the synergistic effect of the 5-HT$_6$ receptor antagonist tested and diazepam cannot be excluded. On the other hand, Marcos *et al.* (2006) demonstrated that the GABA$_A$ receptor antagonist bicuculline did not alter the acetylcholine release produced by SB-357134, another 5-HT$_6$ receptor antagonist. Further studies are necessary to explain this interaction.

The hippocampus seems to be an important site of action of anxiolytic compounds with diverse mechanisms. For example, 5-HT$_{1A}$ receptor partial agonists, benzodiazepine receptor agonists, and ligands of ionotropic or metabotropic glutamatergic receptors exhibit anxiolytic-like effects in various models of anxiety after administration in the hippocampus (Chojnacka-Wojcik *et al.*, 2001; Millan, 2003). Similarly, a 5-HT$_6$ receptor blocker SB-258585, injected into CA1 region of rat hippocampus, produced anxiolytic-like activity in the conflict drinking test (Wesołowska *et al.*, 2007) (Table II). Its effect seems to be specific, as this agent, given at an anxiolytic dose, affected neither the shock threshold nor unpunished water consumption. However, its effect observed in the Vogel test in rats was weaker than that of diazepam.

IV. Prospects for 5-HT$_6$ Receptor Ligands

The above-cited results support the contention that both kinds of 5-HT$_6$ receptor ligands can produce potential antidepressant and anxiolytic activity. It is possible that 5-HT$_6$ receptor agonists and antagonists happen to evoke similar behavioral effects but through different mechanism, although each of them seems to be primarily connected with stimulation or blockade, respectively, of 5-HT$_6$ receptors, given that studied compounds are selective 5-HT$_6$ receptor ligands (Bromidge *et al.*, 1999; Hirst *et al.*, 2003b, 2006; Routledge *et al.*, 2000; Schechter *et al.*, 2008). Second, different neurochemical mechanisms or actions mediated in different brain regions can provide explanations for similar antidepressant/anxiolytic effects of agonists and antagonists.

Indeed, the stimulation of 5-HT$_6$ receptors may evoke antidepressant-like activity, and 5-HT$_6$ receptors seem to play a partial role in the potential antidepressant effect of fluoxetine. The ability of a 5-HT$_6$ receptor agonist to enhance extracellular GABA levels and decrease stimulated glutamatergic neurotransmission supports the hypothesis that 5-HT$_6$ receptor agonists may be effective agents for the treatment of depression, particularly when glutamate levels are enhanced under pathologic circumstances. In contrast, an acute and selective blockade of 5-HT$_6$ receptors may evoke an antidepressant-like effect that does not seem to be dependent on serotonergic neurotransmission but could predominantly involve brain noradrenergic and/or dopaminergic neurotransmission. Moreover, this explanation is supported by results showing that the selective blockade of 5-HT$_6$ receptors induced by SB-399885 may facilitate the anti-immobility activity of antidepressant drugs whose mechanism of action is connected with norepinephrine/dopamine uptake inhibition or with the inhibition of monoamine oxidase-A. However, in this case, one cannot exclude pharmacokinetic interaction yet. Additionally, hippocampus seems to be one of neuroanatomical sites involved in an antidepressant-like effect of a 5-HT$_6$ receptor antagonist.

The ability of 5-HT$_6$ receptor agonists to produce potential anxiolytic activity and to enhance extracellular GABA levels and decrease stimulated glutamatergic neurotransmission seems to support the hypothesis that 5-HT$_6$ receptor agonists may be effective agents for the treatment of anxiety, since it has been demonstrated that activation of GABA receptors is associated with antianxiety effects (Shiah and Yatham, 1998). On the other hand, a selective

5-HT$_6$ receptor antagonist, when administered peripherally or into the hippocampus, may also evoke an anxiolytic-like effect that can be explained by functional interaction between 5-HT$_6$ receptors and the benzodiazepine system. The selective blockade of 5-HT$_6$ receptors, produced by SB-399885, may also facilitate anticonflict effect of diazepam.

Over the past several years, a 5-HT$_6$ receptor has emerged as a highly interesting molecular target that thoroughly interacts with antidepressant/anxiolytic drugs. The results of preclinical studies support the notion that the development of selective 5-HT$_6$ receptor ligands could substantially improve the treatment of affective disorders if their clinical efficacy could be achieved. Additionally, 5-HT$_6$ receptors are almost exclusively expressed in the central nervous system (Monsma *et al.*, 1993; Ruat *et al.*, 1993), thereby reducing the potential for their peripheral side effects. But in the first place, one should keep in mind that most of the present results concern the effects of 5-HT$_6$ receptor ligands after their acute administration. Elucidation of the full behavioral and neurochemical profile of these compounds following repeated treatment is necessary in order to determine suitable recommendations for their potential use in clinic.

References

Ase, A.R., Reader, T.A., Hen, R., Riad, M., and Descarries, L. (2000). Altered serotonin and dopamine metabolism in the CNS of serotonin 5-HT$_{1A}$ or 5-HT$_{1B}$ receptor knockout mice. *J. Neurochem.* **75**, 2415–2426.

Balerio, G.N., Aso, E., and Maldonado, R. (2006). Role of cannabinoid system in the effects induced by nicotine on anxiety-like behaviour in mice. *Psychopharmacology* **184**, 504–513.

Bechtholt, A.J., Valentino, R.J., and Lucki, I. (2008). Overlapping and distinct brain regions associated with the anxiolytic effects of chlordiazepoxide and chronic fluoxetine. *Neuropsychopharmacology* **33**, 2117–2130.

Beck, C.H. (1995). Acute treatment with antidepressant drugs selectively increases the expression of c-fos in the rat brain. *J. Psychiatry Neurosci.* **20**, 23–25.

Benes, F.M., Burke, R.E., Walsh, J., Berrette, S., Matzilevich, D., Minns, M., and Konradi, C. (2004). Acute amygdalar activation induces an upregulation of multiple monoamine G protein coupled pathways in rat hippocampus. *Mol. Psychiatry* **9**, 932–945.

Boast, C.A., Bernard, P.S., Barbaz, B.S., and Bergen, K.M. (1983). The neuropharmacology of various diazepam antagonists. *Neuropharmacology* **22**, 1511–1521.

Bodnoff, S.R., Suranyl-Cadotte, B., Aitken, D.H., Quirion, R., and Meaney, M.J. (1988). The effects of chronic antidepressant treatment in an animal of anxiety. *Psychopharmacology* **95**, 298–302.

Bodnoff, S.R., Suranyl-Cadotte, B., Quirion, R., and Meaney, M.J. (1989). A comparison of the effects of diazepam versus several typical and atypical anti-depressant drugs in an animal model of anxiety. *Psychopharmacology* **97**, 277–279.

Boess, F.G., Monsma, F.J., Carolo, C., Meyer, V., Rudler, A., Zwingelstein, C., and Sleight, A.J. (1997). Functional and radioligand binding characterization of rat 5-HT$_6$ receptors stably expressed in HEK293 cells. *Neuropharmacology* **36**, 713–720.

Boess, F.G., Riemer, C., Bös, M., Bentley, J., Bourson, A., and Sleight, A.J. (1998). The 5-hydroxytryptamine$_6$ receptor-selective radioligand [^{3}H]Ro 63-0563 labels 5-hydroxytryptamine receptor binding sites in rat and porcine striatum. *Mol. Pharmacol.* **54**, 577–583.

Bonasera, S.J., Chu, H.-M., Brennan, T.J., and Tecott, L.H. (2006). A null mutation of the serotonin 6 receptor alters acute responses to ethanol. *Neuropsychopharmacology* **31**, 1801–1813.

Bondi, C.O., Barrera, G., Lapiz, M.D., Bedard, T., Mahan, A., and Morilak, D.A. (2007). Noradrenergic facilitationof shock-probe defensive burying in lateral septum of rats, and modulation by chronic treatment with desipramine. *Prog. Neuropsychopharmacol. Biol. Psychiatry* **31**, 482–495.

Borsini, F. (1995). Role of the serotonergic system in the forced swimming test. *Neurosci. Biobehav. Rev.* **19**, 377–395.

Borsini, F., and Meli, A. (1988). Is the forced swimming test a suitable model for revealing antidepressant activity? *Psychopharmacology* **94**, 147–160.

Borsini, F., and Meli, A. (1990). The forced swimming test: its contribution to the understanding of the mechanisms of action of antidepressants. *Adv. Biosci.* **77**, 63–76.

Borsini, F., Stasi, M.A., Minetti, P., Riccioni, T., and Menini, T. (2008). Effect of ST1936, a 5-HT$_6$ ligand, on rodent adenylate cyclase and forced swimming test. *Int. J. Neuropsychopharmacol.* **11**(Suppl. 1); 123–124.

Brioni, J.D., O'Neill, A.B., Kim, D.J., and Decker, M.W. (1993). Nicotinic receptor agonists exhibits anxiolytic-like effects on the elevated plus-maze test. *Eur. J. Pharmacol.* **238**, 1–8.

Bromidge, S.M., Brown, A.M., Clarke, S.E., Dodgson, K., Gager, T., Grassam, H.L., Jeffrey, P.M., Joiner, G.F., King, F.D., Middlemiss, D.N., Moss, S.F., Newman, H., Riley, G., Routledge, C., and Wyman, P. (1999). 5-Chloro-*N*-(4-methoxy-3-piperazin-1-yl-phenyl)-3-methyl-2-benzothiophenesulfonamide (SB-271046): a potent, selective, and orally bioavailable 5-HT$_6$ receptor antagonist. *J. Med. Chem.* **42**, 202–205.

Carr, G. V., Schechter, L. E., and Lucki I. (2010) Antidepressant and anxiolytic effects of selective 5-HT$_6$ receptor agonists in rats. *Psychopharmacology* **Mar 9** (Epub ahead of print).

Chojnacka-Wojcik, E., Kłodzińska, A., and Pilc, A. (2001). Glutamate receptor ligands as anxiolytics. *Curr. Opin. Investig. Drugs* **2**, 1112–1119.

Cole, D.C., Stock, J.R., Lennox, W.J., Bernotas, R.C., Ellingboe, J.W., Boikess, S., Coupet, J., Smith, D.L., Leung, L., Zhang, G.-M., Feng, X., Kelly, M.F., Galante, R., Huang, P., Dawson, L.A., Marquis, K., Resenzweig-Lipson, S., Beyer, C.E., and Schechter, L.E. (2007). Discovery of N_1-(6-chloroimidazo[2,1-*b*][1,3]-thiazole-5-sulfonyl)tryptamine as a potent, selective, and orally active 5-HT$_6$ receptor agonist. *J. Med. Chem.* **50**, 5535–5538.

Coppell, A.L., Pei, Q., and Zetterström, T.S.C. (2003). Bi-phasic change in BDNF gene expression following antidepressant drug treatment. *Neuropharmacology* **44**, 903–910.

Cryan, J.F., Mombereau, C., and Vassout, A. (2005). The tail suspension test as a model for assessing antidepressant activity: review of pharmacological and genetic studies in mice. *Neurosci. Biobehav. Rev.* **29**, 571–625.

Dawson, L.A., Nguyen, H.Q., and Li, P. (2000). *In vivo* effects of the 5-HT$_6$ antagonist SB-271046 on striatal and frontal cortex extracellular concentrations of noradrenaline, dopamine, 5-HT, glutamate and aspartate. *Br. J. Pharmacol.* **130**, 23–26.

Dawson, L.A., Nguyen, H.Q., and Li, P. (2001). The 5-HT$_6$ receptor antagonist SB-271046 selectively enhances excitatory neurotransmission in the rat frontal cortex and hippocampus. *Neuropsychopharmacology* **25**, 662–668.

Dawson, L.A., Nguyen, H.Q., and Li, P. (2003). Potentiation of amphetamine-induced changes in dopamine and 5-HT by a 5-HT$_6$ receptor antagonist. *Brain Res. Bull.* **59**, 513–521.

de Foubert, G., Carney, S.L., Robinson, C.S., Destexhe, E.J., Tomlinson, R., Hicks, C.A., Murray, T.K., Gaillard, J.P., Deville, C., Xhenseval, V., Thomas, C.E., O'Neill, M.J., and Zetterström, T.S.C. (2004). Fluoxetine-induced change in rat brain expression of brain-derived neurotrophic factor varies depending on length of treatment. *Neuroscience* **128**, 597–604.

de Foubert, G., O'Neill, M.J., and Zetterström, T.S.C. (2007). Acute onset by 5-HT_6-receptor activation on rat brain-derived neurotrophic factor and activity-regulated cytoskeletal-associated protein mRNA expression. *Neuroscience* **147**, 778–785.

Degroot, A., and Treit, D. (2002). Dorsal and ventral hippocampal cholinergic systems modulate anxiety in the plus-maze and shock-probe tests. *Brain Res.* **949**, 60–70.

De Montis, M.G., Mori, A., Crociani, A., and Grappi, S. (2009). Anti anhedonic activity of the 5-HT_6 agonist ST 1936. *Behav. Pharmacol.* **20**, S57.

Detke, M.J., Rickels, M., and Lucki, I. (1995). Active behaviors in the rat forced swimming test differentially produced by serotonergic and noradrenergic antidepressants. *Psychopharmacology* **121**, 66–72.

Di Chiara, G., Valentini, V., Fenu, S., De Luca, M.A., and Borsini, F. (2009). Reinforcing, rewarding and dopamine-stimulant properties of the 5-HT_6 receptor agonist ST1936: the first member of a new class of psychostimulants? *Behav. Pharmacol.* **20**, S9.

Dulawa, S.C., Holick, K.A., Gundersen, B., and Hen, R. (2004). Effects of chronic fluoxetine in animal models of anxiety and depression. *Neuropsychopharmacology* **29**, 1321–1330.

File, S.E., Pellow, S., and Wilks, L. (1985). The sedative effects of CL 218,872, like those of chlordiazepoxide, are reversed by benzodiazepine antagonists. *Psychopharmacology* **85**, 295–300.

Foley, A.G., Murphy, K.J., Hirst, W.D., Gallagher, H.C., Hagan, J.J., Upton, N., Walsh, F.S., and Regan, C.M. (2004). The 5-HT_6 receptor antagonist SB-271046 reverses scopolamine-disrupted consolidation of a passive avoidance task and ameliorates spatial task deficits in aged rats. *Neuropsychopharmacology* **29**, 93–100.

Frantz, K.J., Hansson, K.J., Stouffer, D.G., and Parsons, L.H. (2002). 5-HT_6 receptor antagonism potentiates the behavioral and neurochemical effects of amphetamine but not cocaine. *Neuropharmacology* **42**, 170–180.

Gaynes, B.N., Rush, A.J., Trivedi, M.H., Wisniewski, S.R., Spencer, D., and Fava, M. (2008). The STAR*D study: treating depression in the real world. *Cleve Clin. J. Med.* **75**, 57–66.

Gerard, C., el Mestikawy, S., Lebrand, C., Adrien, J., Ruat, M., Traiffort, E., Hamon, M., and Martres, M.P. (1996). Quantitative RT-PCR distribution of serotonin 5-HT_6 receptor mRNA in the central nervous system of control or 5,7-dihydroxytryptamine-treated rats. *Synapse* **23**, 164–173.

Gerard, C., Martres, M.P., Lefevre, K., Miquel, M.C., Verge, D., Lanfumey, L., Doucet, E., Hamon, M., and el Mestikawy, S. (1997). Immuno-localization of serotonin 5-HT_6 receptor-like material in the rat central nervous system. *Brain Res.* **23**, 207–219.

Glennon, R.A., Lee, M., Rangisetty, J.B., Dukat, M., Roth, B.L., Savage, J.E., McBride, A., Rauser, L., Hufeisen, S., and Lee, D.K. (2000). 2-Substituted tryptamines: agents with selectivity for 5-$HT_{(6)}$ serotonin receptors. *J. Med. Chem.* **43**, 1011–1018.

Hamon, M., Doucet, E., Lefevre, K., Miquel, M.-C., Lanfumey, L., Insausti, R., Frechilla, D., Del Rio, J., and Verge, D. (1999). Antibodies and antisense oligonucleotide for probing the distribution and putative functions of central 5-HT_6 receptors. *Neuropsychopharmacology* **21**, 68–76.

Hensler, J.G. (2006). Serotonergic modulation of the limbic system. *Neurosci. Biobehav. Rev.* **30**, 203–214.

Hirano, K., Piers, T.M., Searle, K.L., Miller, N.D., Rutter, R., and Chapman, P.F. (2009). Procognitive 5-HT_6 antagonists in the rat forced swimming test: Potential therapeutic utility in mood disorders associated with Alzheimer's disease. *Life Sciences* **84**, 558–562.

Hirst, W.D., Abrahamsen, B., Blaney, F.E., Calver, A.R., Aloj, L., Price, G.W., and Medhurst, A.D. (2003a). Differences in the central nervous system distribution and pharmacology of the mouse 5-hydroxytryptamine-6 receptor compared with rat and human receptors investigated by radioligand binding, site-directed mutagenesis, and molecular modeling. *Mol. Pharmacol.* **64**, 1295–1308.

Hirst, W.D., Minton, J.A.L., Bromidge, S.M., Moss, S.F., Latter, A.J., Riley, G., Routledge, C., Middlemiss, D.N., and Price, G.W. (2000). Characterization of [^{125}I]-SB-258585 binding to human recombinant and native 5-HT_6 receptors in rat, pig and human brain tissue. *Br. J. Pharmacol.* **130**, 1597–1605.

Hirst, W.D., Minton, J.A.L., Bromidge, S.M., Routledge, C., Middlemiss, D.N., and Price, G.W. (1999). [^{125}I]-SB-258585 – a selective antagonist radioligand for 5-HT$_6$ receptors. *Br. J. Pharmacol.* **127**, 23P.

Hirst, W.D., Moss, S.F., Bromidge, S.M., Riley, G., Stean, T.O., Rogers, D.C., Sunter, D., Lacroix, L. P., Atkins, A.R., Dawson, L.A., and Upton, N. (2003b). Characterisation of SB-399885, a potent and selective 5-HT$_6$ receptor antagonist. *Society of Neuroscience Meeting Abstr* **576**, 7.

Hirst, W.D., Stean, T.O., Rogers, D.C., Sunter, D., Pugh, P., Moss, S.F., Bromidge, S.M., Riley, G., Smith, D.R., Bartlett, S., Heidbreder, C.A., Atkins, A.R., Lacroix, L.P., Dawson, L.A., Foley, A.G., Regan, C.M., and Upton, N. (2006). SB-399885 is a potent, selective 5-HT$_6$ receptor antagonist with cognitive enhancing properties in aged rat water maze and novel object recognition models. *Eur. J. Pharmacol.* **553**, 109–119.

Hogg, S., and Dalvi, A. (2004). Acceleration of onset of action in schedule-induced polydipsia: combinations of SSRI and 5-HT$_{1A}$ and 5-HT$_{1B}$ receptor agonists. *Pharmacol. Biochem. Behav.* **77**, 69–75.

Horowitz, J.M., Goyal, A., Ramdeen, N., Hallas, B.H., Horowitz, A.T., and Torres, G. (2003). Characterization of fluoxetine plus olanzapine treatment in rats: a behavior, endocrine, and immediate-early gene expression analysis. *Synapse* **50**, 353–364.

Huynh, N.H., and McIntyre, R.S. (2008). What are the implications of the STAR*D trial for primary care? A review and synthesis. *J. Clin. Psychiatry* **10**, 91–96.

Janowsky, D.S., el-Yousef, M.K., Davis, J.M., and Sekerke, H.J. (1972). A cholinergic-adrenergic hypothesis of mania and depression. *Lancet* **2**, 632–635.

Jastrzębska-Więsek, M., Siwek, A., Partyka, A., Mierzejewski, P., Bieńkowski, P., Kołaczkowski, M., and Wesołowska, A. (2010). Potential antidepressant-like effect of the potent serotonin 5-HT$_6$ receptor agonist EMD 386088 in the forced swim test in rats. *Pharmacol. Rep.* **62**(Suppl); 45.

Jayarajan, P., Vishwakarma, S. L., Pendharkar, V. V., Sharma, D. C., Shirsath, V. S., and Nirogi, R. V. S. (2005). Effect of selective 5-HT$_6$ receptor antagonists in rodent models of learning. Society of Neuroscience Meeting Abstract 881.8.

Kohen, R., Fashingbauer, L.A., Heidmann, D.E., Guthrie, C.R., and Hamblin, M.W. (2001). Cloning of the mouse 5-HT$_6$ serotonin receptor and mutagenesis studies of the third cytoplasmatic loop. *Brain Res. Mol. Brain Res.* **90**, 110–117.

Kostowski, W. (1985). Possible relationship of the locus coeruleus-hippocampal noradrenergic neurons to depression and mode of action of antidepressant drugs. *Polish J. Pharmacol. Pharm.* **37**, 727–743.

Kramlinger, K.G., Petersen, G.C., Watson, P.K., and Leonard, L.L. (1985). Metyrapone for depression and delirium secondary to Cushing's syndrome. *Psychosomatics* **26**, 67–71.

Lacroix, L.P., Dawson, L.A., Hagan, J.J., and Heidbreder, C.A. (2004). 5-HT$_6$ receptor antagonist SB-271046 enhances extracellular levels of monoamines in the rat medial prefrontal cortex. *Synapse* **51**, 158–164.

Lange, C., and Irle, E. (2004). Enlarged amygdala volume and reduced hippocampal volume in young women with major depression. *Psychol. Med.* **324**, 1059–1064.

Leung, J.W.C., and Xue, H. (2003). GABAergic functions and depression: from clinical therapies to herbal medicine. *Curr. Drug Targets CNS Neurol. Disorders* **2**, 363–373.

Li, Z., Huang, M., Prus, A.J., Dai, J., and Meltzer, H.Y. (2007). 5-HT$_6$ receptor antagonist SB-399885 potentiates haloperidol and risperidone-induced dopamine efflux in the medial prefrontal cortex or hippocampus. *Brain Res.* **1134**, 70–78.

Lieben, C.K., Blokland, A., Sik, A., Sung, E., van Nieuwenhuizen, P., and Schreiber, R. (2005). The selective 5-HT$_6$ receptor antagonist Ro4368554 restores memory performance in cholinergic and serotonergic models of memory deficiency in the rat. *Neuropsychopharmacology* **30**, 2169–2179.

Liljequist, S., and Engel, J.A. (1984). The effects of GABA and benzodiazepine receptor antagonists on the anti-conflict actions of diazepam or ethanol. *Pharmacol. Biochem. Behav.* **21**, 521–525.

Loiseau, F., Dekeyne, A., and Millan, M.J. (2008). Pro-cognitive effects of 5-HT_6 receptor antagonists in the social recognition procedure in rats: implication of the frontal cortex. *Psychopharmacology* **196**, 93–104.

Malberg, J.E., and Schechter, L.E. (2005). Increasing hippocampal neurogenesis: a novel mechanism for antidepressant drugs. *Curr. Pharm. Des.* **11**, 145–155.

Marcos, B., Gil-Bea, F.J., Hirst, W.D., Garcia-Alloza, M., and Ramirez, M.J. (2006). Lack of localization of 5-HT_6 receptors on cholinergic neurons: implication of multiple neurotransmitter systems in 5-HT_6 receptor-mediated acetylcholine release. *Eur. J. Neurosci.* **24**, 1299–1306.

Mattsson, C., Sonesson, C., Sandahl, A., Greiner, H.E., Gassen, M., Plaschke, J., Leibrock, J., and Böttcher, H. (2005). 2-Alkyl-3-(1,2,3,6-tetrahydropyridin-4-yl)-1*H*-indoles as novel 5-HT_6 receptor agonists. *Bioorg. Med. Chem. Lett.* **15**, 4230–4234.

Millan, M.J. (2003). The neurobiology and control of anxious states. *Prog. Neurobiol.* **70**, 83–244.

Monsma, F.J., Shen, Y., Ward, R.P., Hamblin, M.W., and Sibley, D.R. (1993). Cloning and expression of a novel serotonin receptor with high affinity for tricyclic psychotropic drugs. *Mol. Pharmacol.* **43**, 320–327.

Murphy, B.E., Dhar, V., Ghadirian, A.M., Chouinard, G., and Keller, R. (1991). Response to steroid suppression in major depression resistant to antidepressant therapy. *J. Clin. Psychopharmacol.* **11**, 121–126.

Nestler, E.J., Barrot, M., DiLeone, R.J., Eisch, A.J., Gold, S.J., and Monteggia, L.M. (2002). Neurobiology of depression. *Neuron* **34**, 13–25.

Neumeister, A., Wood, S., Bonne, O., Nugent, A.C., Luckenbaugh, D.A., Young, T., Bain, E.E., Charney, D.S., and Drevets, W.C. (2005). Reduced hippocampal volume in unmedicated, remitted patients with major depression versus control subjects. *Biol. Psychiatry* **57**, 935–937.

O'Brien, J.T., Lloyd, A., McKeith, I., Gholkar, A., and Ferrier, N. (2004). A longitudinal study of hippocampal volume, cortisol levels, and cognition in older depressed subjects. *Am. J. Psychiatry* **161**, 2081–2090.

Otano, A., Frechilla, D., Cobreros, A., Cruz-Orive, L.M., Insausti, A., Insausti, R., Hamon, M., and Del Rio, J. (1999). Anxiogenic-like effects and reduced stereological counting of immunolabelled 5-hydroxytryptamine$_6$ receptors in rat nucleus accumbens by antisense oligonucleotides. *Neuroscience* **92**, 1001–1009.

Paul, I.A., and Skolnick, P. (2003). Glutamate and depression: clinical and preclinical studies. *Ann. N. Y. Acad. Sci.* **1003**, 250–272.

Pei, Q., Zetterström, T.S.C., Sprakes, M.E., Tordera, R., and Sharp, T. (2003). Antidepressant drug treatment induces Arc gene expression in the rat brain. *Neuroscience* **121**, 975–982.

Pierce, P.A., Xie, G.-X., Levine, J.D., and Peroutka, S.J. (1996). 5-Hydroxytryptamine receptor subtype messenger RNAs in rat peripheral sensory and sympathetic ganglia: a polymerase chain reaction study. *Neuroscience* **70**, 553–559.

Popik, P., Kozela, E., and Krawczyk, M. (2003). Nicotine and nicotinic receptor antagonists potentiate the antidepressant-like effects of imipramine and citalopram. *Br. J. Pharmacol.* **139**, 1196–1202.

Porsolt, R.D., Bertin, A., Blavet, N., Deniel, M., and Jalfre, M. (1979). Immobility induced by forced swimming in rats: effects of agents which modify central catecholamine and serotonin activity. *Eur. J. Pharmacol.* **57**, 201–210.

Przegaliński, E., Tatarczyńska, E., Dereń-Wesołek, A., and Chojnacka-Wojcik, E. (1997). Antidepressant-like effects of a partial agonist at strychnine-insensitive glycine receptors and a competitive NMDA receptor antagonist. *Neuropharmacology* **36**, 31–37.

Pullagurla, M., Bondareva, T., Young, R., and Glennon, R.A. (2004). Modulation of the stimulus effects of (+)amphetamine by the 5-HT_6 antagonist MS-245. *Pharmacol. Biochem. Behav.* **78**, 263–268.

Riemer, C., Borroni, E., Levet-Trafit, B., Martin, J.R., Poli, S., Porter, R.H., and Bos, M. (2003). Influence of the 5-HT_6 receptor on acetylcholine release in the cortex: pharmacological characterization of 4-(2-bromo-6-pyrrolidin-1-ylpyridine-4-sulfonyl)phenylamine, a potent and selective 5-HT_6 receptor antagonist. *J. Med. Chem.* **46**, 1273–1276.

Routledge, C., Bromidge, S.M., Moss, S.F., Price, G.W., Hirst, W., Newman, H., Riley, G., Gager, T., Stean, T., Upton, N., Clarke, S.E., Brown, A.M., and Middlemiss, D.N. (2000). Characterisation of SB-271046-A: a potent, selective and orally active 5-HT_6 receptor antagonist. *Br. J. Pharmacol.* **130**, 1606–1612.

Ruat, M., Traiffort, E., Arrang, J.M., Tardivel-Lacombe, J., Diaz, J., Leurs, R., and Schwartz, J.C. (1993). A novel rat serotonin (5-HT_6) receptor: molecular cloning, localization and stimulation of cAMP accumulation. *Biochem. Biophys. Res. Commun.* **193**, 268–276.

Rush, A.J., Trivedi, M.H., Wisniewski, S.R., Nierenberg, A.A., Stewart, J.W., Warden, D., Niederehe, G., Thase, M.E., Lavori, P.W., Lebowitz, B.D., McGrath, P.J., Rosenbaum, J.F., Sackeim, H.A., Kupfer, D.J., Luther, J., and Fava, M. (2006). Acute and longer-term outcomes in depressed outpatients requiring one or several treatment steps: a STAR*D report. *Am. J. Psychiatry* **163**, 1905–1917.

Russo-Neustadt, A.A., and Chen, M.J. (2005). Brain-derived neurotrophic factor and antidepressant activity. *Curr. Pharm. Des.* **11**, 1495–1510.

Sanacora, G., Gueorguieva, R., Epperson, C.N., Wu, Y.T., Appel, M., Rothman, D.L., Krystal, J.H., and Mason, G.F. (2004). Subtype-specific alterations of γ-aminobutyric acid and glutamate in patients with major depression. *Arch. Gen. Psychiatry* **61**, 705–713.

Sanacora, G., Mason, G.F., and Krystal, J.H. (2000). Impairment of GABAergic function in depression: new insights from neuroimaging. *Crit. Rev. Neurobiol.* **14**, 23–45.

Sarter, M., and Parikh, V. (2005). Choline transporters, cholinergic transmission and cognition. *Nat. Rev. Neurosci.* **6**, 48–56.

Saylam, C., Ucerler, H., Kitis, O., Ozand, E., and Gonul, A.S. (2006). Reduced hippocampal volume in drug-free depressed patients. *Surg. Radiol. Anat.* **28**, 82–87.

Schechter, L.E., Lin, Q., Smith, D.L., Zhang, G., Shan, Q., Platt, B., Brandt, M.R., Dawson, L.A., Cole, D., Bernotas, R., Robichaud, A., Resenzweig-Lipson, S., and Beyer, C.E. (2008). Neuropharmacological profile of novel and selective 5-HT_6 receptor agonists: WAY-181187 and WAY-208466. *Neuropsychopharmacology* **33**, 1323–1335.

Sebben, M., Ansanay, H., Bockaert, J., and Dumuis, A. (1994). 5-HT_6 receptors positively coupled to adenylyl cyclase in striatal neurons in culture. *NeuroReport* **5**, 2553–2557.

Sherman, A.D., and Allers, G.L. (1980). Relationship between regional distribution of imipramine and its effect on learned helplessness in the rat. *Neuropharmacology* **19**, 159–162.

Shiah, I.S., and Yatham, L.N. (1998). GABA function in mood disorders; an update and critical review. *Life Sci.* **63**, 1289–1303.

Shimizu, E., Hashimoto, K., Okamura, N., Koike, K., Komatsu, N., Kumakiri, C., Nakazato, M., Watanabe, H., Shinoda, N., Okada, S., and Iyo, M. (2003). Alterations of serum levels of brain-derived neurotrophic factor (BDNF) in depressed patients with or without antidepressants. *Biol. Psychiatry* **54**, 70–75.

Shirayama, Y., Chen, A.C., Nakagawa, S., Russell, D.S., and Duman, R.S. (2002). Brain-derived neurotrophic factor produces antidepressant effects in behavioural models of depression. *J. Neurosci.* **22**, 3251–3261.

Shirazi-Southall, S., Rodriguez, D.E., and Nomikkos, G.G. (2002). Effects of typical and atypical antipsychotics and receptor selective compounds on acetylcholine efflux in the hippocampus of the rat. *Neuropsychopharmacology* **26**, 583–594.

Shytle, R.D., Silver, A.A., Lukas, R.J., Newman, M.B., Sheehan, D.V., and Sanberg, P.R. (2002). Nicotinic acetylcholine receptors as targets for antidepressants. *Mol. Psychiatry* **7**, 525–535.

Svenningsson, P., Tzavara, E.T., Qi, H., Carruthers, R., Witkins, J.M., Nomikos, G.G., and Greengard, P. (2007). Biochemical and behavioral evidence for antidepressant-like effects of 5-HT_6 receptor stimulation. *J. Neurosci.* **27**, 4201–4209.

Treit, D., Pinel, J.P., and Fibiger, H.C. (1981). Conditioned defensive burying: a new paradigm for the study of anxiolytic agents. *Pharmacol. Biochem. Behav.* **15**, 619–626.

Videbech, P., and Ravnkilde, B. (2004). Hippocampal volume and depression: a meta-analysis of MRI studies. *Am. J. Psychiatry* **161**, 1957–1966.

Vogt, I.R., Shimron-Abarbanell, D., Neidt, H., Erdmann, J., Cichon, S., and Schulze, T.G. (2000). Investigation of the human serotonin 6 [5-HT_6] receptor gene in bipolar affective disorder and schizophrenia. *Am. J. Med. Genet.* **96**, 217–221.

Vythilingam, M., Vermetten, E., Anderson, G.M., Luckenbaugh, D., Anderson, E.R., Snow, J., Staib, L.H., Charney, D.S., and Bremner, J.D. (2004). Hippocampal volume, memory, and cortisol status in major depressive disorder: effects of treatment. *Biol. Psychiatry* **56**, 101–112.

Ward, R.P., and Dorsa, D.M. (1996). Colocalization of serotonin receptor subtypes 5-HT_{2A}, 5-HT_{2C}, and 5-HT_6 with neuropeptides in rat striatum. *J. Comp. Neurol.* **370**, 405–414.

Ward, R.P., Hamblin, M.W., Lachowicz, J.E., Hoffman, B.J., Sibley, D.R., and Dorsa, D.M. (1995). Localization of serotonin subtype 6 receptor messenger RNA in the rat brain by in situ hybridization histochemistry. *Neuroscience* **64**, 1105–1111.

Wesołowska, A. (2007). Study into a possible mechanism responsible for the antidepressant-like activity of the selective 5-HT_6 receptor antagonist SB-399885 in rats. *Pharmacol. Rep.* **59**, 664–671.

Wesołowska, A. (2008). The anxiolytic-like effect of the selective 5-HT_6 receptor antagonist SB-399885: the impact of benzodiazepine receptors. *Eur. J. Pharmacol.* **580**, 260–355.

Wesołowska, A., and Nikiforuk, A. (2007). Effects of the brain-penetrant and selective 5-HT_6 receptor antagonist SB-399885 in animal models of anxiety and depression. *Neuropharmacology* **52**, 1274–1283.

Wesołowska, A., and Nikiforuk, A. (2008). The selective 5-HT_6 receptor antagonist SB-399885 enhances anti-immobility action of antidepressants in rats. *Eur. J. Pharmacol.* **582**, 88–93.

Wesołowska, A., Nikiforuk, A., and Stachowicz, K. (2007). Anxiolytic-like and antidepressant-like effects produced by the selective 5-HT_6 receptor antagonist SB-258585 after intrahippocampal administration to rats. *Behav. Pharmacol.* **18**, 439–446.

Winter, H., and Irle, E. (2004). Hippocampal volume in adult burn patients with and without post-traumatic stress disorder. *Am. J. Psychiatry* **16**, 2194–2200.

Woods, A., Smith, C., Szewczak, M., Dunn, R.W., Cornfeldt, M., and Corbett, R. (1993). Selective serotonin re-uptake inhibitors decrease schedule-induced polydipsia in rats: a potential model for obsessive compulsive disorder. *Psychopharmacology* **112**, 195–198.

Woolley, M.L., Marsden, C.A., and Fone, K.C.F. (2004). 5-HT_6 receptors. *Curr. Drug Targets CNS Neurol. Disorders* **3**, 59–79.

Yau, J.L.W., Noble, J., Widdowson, J., and Seckl, J.R. (1997). Impact of adrenalectomy on 5-HT_6 and 5-HT_7 receptor gene expression in the rat hippocampus. *Mol. Brain Res.* **45**, 182–186.

Yilmaz, A., Schulz, D., Aksoy, A., and Canbeyli, R. (2002). Prolonged effect of an anesthetic dose of ketamine on behavioral despair. *Pharmacol. Biochem. Behav.* **71**, 349–352.

Yoshioka, M., Matsumoto, M., Togashi, H., Mori, K., and Saito, H. (1998). Central distribution and function of 5-HT_6 receptor subtype in the rat brain. *Life Sci.* **62**, 1473–1477.

Zarate, C.A., Singh, J.B., Carlson, P.J., Brutsche, N.E., Ameli, R., Luckenbaugh, M.A., Charney, D.S., and Manji, H.K. (2006). A randomized trial of an *N*-methyl-D-aspartate antagonist in treatment-resistant major depression. *Arch. Gen. Psychiatry* **63**, 856–864.

THE 5-HT_6 RECEPTOR AS A TARGET FOR DEVELOPING NOVEL ANTIOBESITY DRUGS

David Heal, Jane Gosden and Sharon Smith

RenaSci Consultancy Limited, BioCity, Nottingham NG1 1GF, UK

I. Introduction

It is almost impossible for a single week to pass without hearing another horror story about the increasing prevalence of obesity and the global epidemic of disease and death that will follow in its wake. The epidemiological data supporting this view are compelling (Brown *et al.*, 2010; Centers for Disease Control and Prevention, 2010; Flegal *et al.*, 2010; International Obesity Task Force, 2010) and there can be no question that the high prevalence of obesity is a modern phenomenon that has grown to epidemic proportions in little over three decades. Even as late as the 1980s, obesity in European countries was at a relatively low level (Fig. 1). However in the intervening years, the prevalence of obesity across the whole of Europe has increased steadily to the extent that in many countries, including the UK, ~25% of adult males and females meet the criteria for a clinical diagnosis of obesity (body mass index [BMI; height in m^2/weight in kg] of 30 or greater).

There is a good consensus that the major factors leading to increased obesity are easy access to a "fast food" diet that is too rich in fat, sugar, and salt combined with a much more sedentary lifestyle (Deedwania, 2004; Ello-Martin *et al.*, 2005; Prentice and Jebb, 1995). However, recent epidemiological data in children

INTERNATIONAL REVIEW OF
NEUROBIOLOGY, VOL. 96
DOI: 10.1016/B978-0-12-385902-0.00004-8

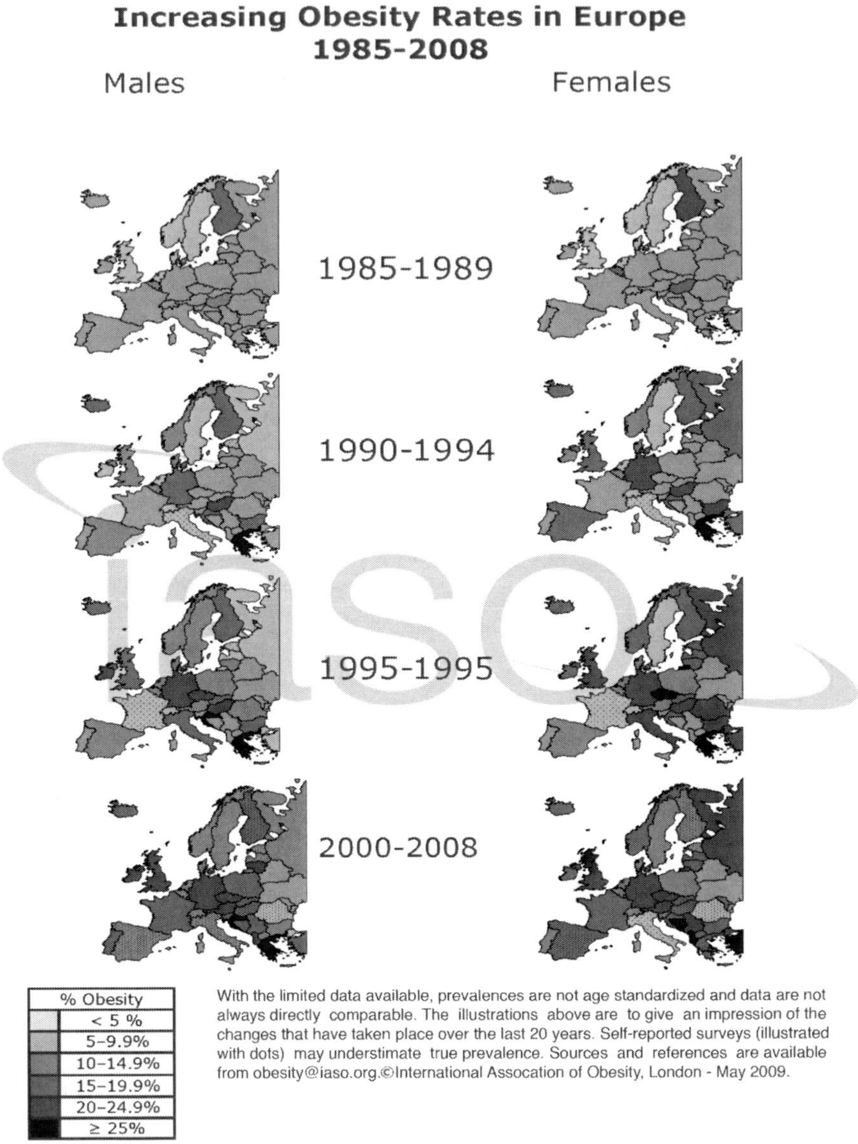

FIG. 1. Prevalence rates and trends in obesity in Europe 1985–2008. Reproduced with permission from IASO. (For interpretation of the references to color in this figure legend, the reader is referred to the Web version of this chapter.).

indicate that the amount of exercise taken has relatively little influence on the occurrence and degree of obesity (Metcalf *et al.*, 2010), whereas there is a very strong association between daily calorie intake, consumption of "fast food," and obesity (Block *et al.*, 2004; Bowman *et al.*, 2004). Although rates of adult obesity may now be reaching a plateau, obesity rates in adolescents are still increasing (CDC, 2010). This generation of children is far less healthy than its predecessors and some experts have even speculated that todays children may be the first generation to have a shorter life-expectancy than their parents (Daniels, 2006). Obesity is not exclusively a problem for wealthy, industrialized countries, although it is undoubtedly where the epidemic started. The export of a fast food diet and Western culture to developing nations has turned it into a global problem. There is now the conundrum that obesity and food poverty are simultaneous health problems in many third world countries (IOTF, 2010).

The problem of obesity is not merely a cosmetic issue, even though it is strongly associated with low self-esteem and a higher incidence of affective disorders (Ackard *et al.*, 2003; Goodman and Whitaker, 2002; Simon *et al.*, 2006; Wardle *et al.*, 2002). The metabolic dysregulation present in obesity predisposes the person to developing dyslipidemia (elevated plasma concentration of triglycerides and low-density cholesterol [LDL-C] and decreased plasma concentrations of high-density cholesterol [HDL-C]), hypertension, proinflammatory atherogenesis, prediabetes (i.e., insulin resistance and impaired glucose tolerance), and Type 2 diabetes (Mokdad *et al.*, 2001; National Cholesterol Education Program Adult Treatment Panel 3 [NCEP–ATP 3], 2002; Pi-Sunyer, 2002). All of these diseases are established risk factors for increased cardiovascular morbidity and mortality (Beckman *et al.*, 2002; Flack *et al.*, 2003; Law *et al.*, 2003; Montani *et al.*, 2002; NCEP–ATP 3, 2002; Rashid *et al.*, 2003; UK Prospective Diabetes Study Group UKPDS 38, 1998; UKPDS 66, 2004). In addition to being a predisposing factor for cardiovascular disease, obesity has also been cited as causal in $\leq$30% of cases of cancer of the colon, breast, kidney and digestive tract (IOTF, 2010) and in long-term debilitating diseases such as sleep apnea, aggravated arthritis, gout and gallstones (Bhole *et al.*, 2010; Felson, 1996; Heshka and Heymsfield, 2001; Wolk *et al.*, 2003). The increased mortality risk of obesity follows a J-shaped curve and while an approximate doubling of risk does not occur until BMI exceeds 40 (designated severe or morbid obesity), when the prevalence of obesity is added into the equation, the medical consequences of the epidemic could potentially overwhelm the healthcare systems in many countries. Although for Caucasians the risk of developing comorbid diseases such as Type 2 diabetes increases substantially at a BMI of $\geq$30, for many non-Caucasian races, including Asians and Hispanics who tend to develop the more dangerous abdominal form of obesity, their chances of developing these life-threatening metabolic diseases occur at much lower BMI values (CDC, 2009; Deutenberg *et al.*, 2002; James *et al.*, 2002; Moon *et al.*, 2002).

The inexorable and rapid growth in the number of people with obesity is a testament to the conclusion that adherence to a balanced diet and healthy lifestyle has failed to address this problem for a significant minority of subjects. It is the reason why for many obese subjects, diet and behavioral modification will need to be supplemented by pharmacotherapy, or increasingly, bariatric surgery. A list of marketed antiobesity drugs and drug candidates in clinical development is shown in Table I.

Currently, there are two prescription drugs that have been approved for the long-term treatment of obesity, that is, orlistat and sibutramine. Orlistat is a gastric lipase inhibitor that prevents the absorption from the gut of ~30% of the calories consumed as fat (McNeely and Benfield, 1998). Sibutramine is a centrally acting drug that inhibits the reuptake of 5-hydroxytryptamine (5-HT or serotonin) and noradrenaline (norepinephrine). Sibutramine reduces body weight by decreasing the amount of calories consumed by enhancing satiety (the feeling of fullness that terminates a meal) and increasing energy expenditure via thermogenesis (Connoley *et al.*, 1999; Halford *et al.*, 2010; Heal *et al.*, 1998; Saraç *et al.*, 2006). Although the weight-loss benefits of both drugs have been demonstrated through improvements in a range of cardiometabolic risk factors, for example, reductions in visceral adiposity and plasma triglycerides, together with increases in insulin sensitivity and plasma HDL-C (de Castro *et al.*, 2009; Dujovne *et al.*, 2001; Florentin *et al.*, 2009; Fujioka *et al.*, 2000; Gokcel *et al.*, 2001; Heymsfield *et al.*, 2000; Hollander *et al.*, 1998; Kelley *et al.*, 2002; McNulty *et al.*, 2003; Richelsen *et al.*, 2007; Torgerson *et al.*, 2004), for most subjects, neither drug delivers the double-digit reduction in body weight that is desired by physicians and their patients (Dutton *et al.*, 2010; Haddock *et al.*, 2002; Li *et al.*, 2005; Linné *et al.*, 2002). In addition to limited efficacy, orlistat and sibutramine have side effects that derive from their respective pharmacological actions. By impairing absorption of fat from the gut, orlistat produces gastrointestinal disturbances that are unpleasant and embarrassing, but are not dangerous. Sibutramine's action to inhibit noradrenaline and adrenaline reuptake produces mean increases of 3–5 mmHg in blood pressure and 5–7 bpm in heart rate. It was these cardiovascular side effects that prompted the Sibutramine Cardiovascular OUTcome (SCOUT) trial to determine the long-term effects of weight loss produced by sibutramine on mortality in a patient population at high cardiovascular risk. This "enriched" patient population was selected because it increased the probability of achieving a definitive outcome at the end of the trial. The flaw in the design is that, according to its Product Label, sibutramine is not recommended for use in most of these patients. It was the subsequent finding that mortality was increased by ~10% in the sibutramine treatment group (Food and Drugs Administration [FDA], 2009, 2010) which resulted in its suspension in Europe by European Medicines Agency (EMA) (2010), and in the USA, a review of its safety by the FDA (2009, 2010). Although the future status

Table I
ANTIOBESITY DRUGS AND DRUG CANDIDATES.

Drug	Trade Name	Mode of Action	Company	USA	Territory Not Specified	European Union
Orlistat	Xenical	Lipase inhibitor	Roche	Marketed		Marketed
Orlistat	Alli (OTC)	Lipase inhibitor	Roche	Marketed		Marketed
Sibutramine	Reductil, Meridia	Noradrenaline (NA) + 5-HT reuptake inhibitor	Abbott	Under review		Suspended
Phentermine	Ionamin, Duromine	NA + dopamine (DA) releasing agent	Generic	Marketed		Withdrawn
Diethylpropion	Tenuate, Apisate	Sympathomimetic	Generic	Marketed		Withdrawn
Topiramate/phentermine	Qnexa	Unknown/NA + DA releasing agent	Vivus	Preregistration		Phase 3
Lorcaserin (APD356)	Lorqess	5-HT_{2C} agonist	Arena/Eisai	Preregistration		Phase 3
Bupropion/naltrexone	Contrave	DA reuptake inhibitor/ opioid antagonist	Orexigen/Takeda	Preregistration		Phase 3
Liraglutide	Victoza	GLP-1 agonist	Novo-Nordisk		Phase 3	
ATL962	Cetilistat	Lipase inhibitor	Takeda	Phase 3		Phase 2
TM30338	Obinepitide	NPY Y_2/Y_4 agonist	7-TM		Phase 2	
TM30339	–	Y_2 agonist	7-TM		Phase 2	
S2367	Velneperit	Y_5 antagonist	Shinogi		Phase 2	
NS2330	Tesofensine	Triple monoamine reuptake inhibitor	Neurosearch		Phase 2	
PRX00933	–	5-HT_{2C} agonist	Proximagen		Phase 2	
Zonisamide + bupropion	Empatic	Unknown/DA reuptake inhibitor	Orexigen		Phase 2	

(continued)

Table I (*continued*)

Drug	Trade Name	Mode of Action	Company	USA	Territory Not Specified	European Union
MKD493	–	MC_4 agonist	Merck		Phase 2	
Symlin + metreleptin	–	Amylin agonist + leptin agonist	Amylin/Takeda		Phase 2	
ALB127158	–	MCH_1 antagonist	Albany		Phase 1	
Peptide YY (PYY)	–	Y_2 agonist	Amylin		Phase 1	
S234462	–	Y_5 antagonist	Shinogi		Phase 1	
Betahistine	Histalaen	H_1 agonist/H_3 antagonist	Obecure		Phase 1	
GSK1521498	–	μ-Opioid inverse agonist	GSK		Phase 1	
ZYO1	–	CB_1 antagonist	Zydus		Phase 1	
TM38837	–	Peripheral; CB_1 receptor antagonist	7-TM		Phase 1	
ZYOG1	–	GLP1 agonist	Zydus		Phase 1	
ZGN433	–	Not revealed	Zafgen		Phase 1	
NN9161	–	Not revealed	Novo-Nordisk		Phase 1	
OAP189	–	Not revealed	Pfizer		Phase 1	
PF2575799	–	Not revealed	Pfizer		Phase 1	
TKS1225	–	Not revealed	Pfizer		Phase 1	
HPP404	–	Not revealed	TransTech		Phase 1	
P120107	–	Not revealed	Piramal		Phase 1	

OTC: Over-the-counter (non-prescription) medicine. *Source*: MedTrack®.

of sibutramine in the USA is uncertain, data released by the FDA revealed that the use of sibutramine in the SCOUT trial was not associated with increased mortality in subjects who did not have preexisting cardiovascular disease, that is, the appropriate patient group for sibutramine treatment. It therefore raises the difficult ethical dilemma of whether manufacturers should be asked to demonstrate the clinical benefits of their drugs by performing "off-label" outcome trials in inappropriate patient populations simply to generate definitive endpoints.

The results from the SCOUT trial are the latest in a long line of safety alerts that have dogged pharmacotherapy in this therapeutic indication. In the 1960s, it was the diversion and abuse of d-amphetamine when it was widely prescribed as a short-term appetite suppressant. In the 1990s, it was the occurrence of primary pulmonary hypertension and cardiac valvulopathy that led to the global withdrawal of fenfluramine and dexfenfluramine. Most recently, the cannabinoid CB_1 receptor antagonist, rimonabant, was approved as an antiobesity drug in Europe only to be withdrawn within 2 years on suspicion that it could cause anxiety, depression and suicidal ideation.

In addition to the antiobesity drugs, orlistat and sibutramine, there are a number of older, sympathomimetics that have been approved as short-term (<12 weeks) weight-loss treatments. In the USA, phentermine is the most widely prescribed antiobesity drug by a long measure. In Europe, however, EMA withdrew the licenses for all of these sympathomimetic drugs on the basis that their long-term benefit was unproven and there were potential safety risks associated with their use (EMA, 1999).

Although the development pipeline for antiobesity drugs is relatively sparse (Table I), there are three drug candidates currently undergoing regulatory evaluation by the FDA, that is, Qnexa (a topirimate/phentermine combination product), lorcaserin (a 5-HT$_{2C}$ receptor agonist), and Contrave (a naltrexone/bupropion combination product). The FDA's Advisory Committee recently reviewed Qnexa and, despite a neutral opinion by the FDA's medical assessor, voted 10-6 against approval based on a range of potential safety concerns. We will have to wait to learn whether the FDA resolves these issues with Vivus Pharmaceuticals and the drug will ultimately receive US marketing authorization. The other two drug candidates are also due to be reviewed by the FDA's Advisory Committee within the next 6 months, and by mid-2011, it should be clear whether or not the current formulary will be supplemented by one or more new antiobesity drugs. None of these drugs is under review by EMA, and consequently, they are not likely to be entrants to the European antiobesity market in the near future. Since the guidance for the approval of antiobesity drugs differs between the USA and the European Union, it is not certain whether these drugs would satisfy European regulatory requirements (EMA, 2007). For example, the primary efficacy endpoint demanded by EMA is a mean weight loss from baseline of $\geq 10\%$. On the basis of current clinical findings, Qnexa appears to be the only one of the three that is sufficiently efficacious to satisfy this criterion (Bays, 2010; Vivus, 2009), while Contrave and lorcaserin appear to fall considerably short of the European primary efficacy endpoint (Greenway *et al.*, 2007; Kim, 2010; Smith *et al.*, 2009, 2010).

Behind these compounds sit the GLP-1 receptor agonist, liraglutide, and the gastric lipase inhibitor, cetilistat. The former has already received US and European approval as an injectable treatment for Type 2 diabetes. The known ability of GLP-1 receptor agonists to reduce body weight in diabetic subjects, a patient group that is notoriously resistant to the actions of antiobesity drugs, has undoubtedly prompted Novo-Nordisk to seek to expand the drug's metabolic applications into the treatment of obesity. Cetilistat was discovered and put into early clinical development by Alizyme, but it has now been acquired by Takeda. Although its mode of action is identical to that of orlistat, unlike the latter it is a conventional small molecule compound. Alizyme claimed that cetilistat is advantaged over orlistat by virtue of having a much reduced potential to cause gastrointestinal adverse events. At earlier stages of clinical development are compounds acting at a wide range of central and peripheral drug targets, including the monoamine, cannabinoid, and opioid neurotransmitter systems, hypothalamic neuropeptides and gut hormones. The 5-HT$_6$ receptor ligands undergoing evaluation as potential novel drugs for the treatment of obesity are shown in Table II. All of these compounds are at the early stages of development, that is, in preclinical development or the research phase.

In this chapter, we will concentrate on the role of 5-HT$_6$ receptors in regulating food intake and body weight and the potential of compounds directed against this receptor as novel drugs to treat obesity and its related comorbidities. Readers

Table II
5-HT$_6$ Receptor Ligands as Drug Development Candidates in Obesity.

Compound	Company	Functional Activity	Status
PRX07034	Epix	Antagonist	Phase 1A Current status not known
E6837	Esteve	Partial agonist	Preclinical
MEM68626	Roche	Antagonist	Preclinical
SUVN504	Suven	Antagonist	Preclinical
SUVN51005	Suven	Antagonist	Preclinical
Not revealed	Merck	Not revealed	Preclinical
Not revealed	Albany Molecular Research	Not revealed	Research
BVT.74316	Biovitrum	Antagonist	Discontinued
PRX07034	Epix	Antagonist	Not known

Source: Company websites and MedTrack®.

wishing to gain a greater insight into the chemistry and structure–activity relationships of various chemical classes of 5-HT$_6$ receptor ligands or to the complexities of antiobesity drug development are directed to the reviews of Heal *et al.* (2008, 2009), respectively.

II. 5-HT$_6$ Receptors in the Central Nervous System

The 5-HT$_6$ receptor is a G-protein-coupled receptor that was first cloned with RT-PCR from rat striatal tissue early in the 1990s (Monsma *et al.*, 1993; Ruat *et al.*, 1993). It is a seven-transmembrane-spanning protein of about 440 amino acids (Kohen *et al.*, 1996). When characterized in transfected cell lines, the 5-HT$_6$ receptor was demonstrated to be positively coupled to adenylyl cyclase (Max *et al.*, 1995; Monsma *et al.*, 1993; Ruat *et al.*, 1993).

The 5-HT$_6$ receptor is a very unusual member of the serotonin receptor family because it is located almost exclusively within the CNS (Monsma *et al.*, 1993; Ruat *et al.*, 1993). In rat brain, the highest levels of receptor mRNA are present in the striatum, olfactory tubercle, nucleus accumbens, hippocampus, cortex, cerebellum, hypothalamus, and amygdala (Monsma *et al.*, 1993; Ruat *et al.*, 1993; Ward *et al.*, 1995). The highest density of mRNA within the hippocampus was found to be in the dentate gyrus, CA1, CA2 and CA3 regions (Ruat *et al.*, 1993; Ward *et al.*, 1995). Localization of 5-HT$_6$ receptor protein has also been determined using immunohistochemical or radiolabeling techniques. Receptor presence in rat brain was found in the olfactory tubercle, cortex, nucleus accumbens, striatum, hippocampus, cerebellum, thalamus, substantia nigra, superficial layer of the superior colliculus, motor trigeminal nucleus, facial nucleus and hypothalamus (Fone *et al.*, 2002; Gerard *et al.*, 1997; Hamon *et al.*, 1999; Roberts *et al.*, 2002). Notably, the hypothalamus is the brain region responsible for regulation of food intake and energy expenditure, and is an important site for the action of centrally-acting antiobesity drugs. Hence, there is a very good correlation between the distributions of 5-HT$_6$ receptor mRNA and protein in rat brain. The human 5-HT$_6$ receptor was cloned by Kohen *et al.* (1996) who found the highest expression of mRNA to be in the caudate nucleus, followed by the hippocampus and amygdala. Low expression levels were found in the thalamus, subthalamic nuclei and substantia nigra. East *et al.* (2002) found that the relative distribution of the radiolabeled 5-HT$_6$ antagonist, SB-258585, in the striatum, hippocampus, and cortex was similar to that reported in the rat. In addition, the distribution was in agreement with 5-HT$_6$ receptor mRNA determined by *in situ* hybridization (East *et al.*, 2002). However, in comparison with rats and humans, levels of both 5-HT$_6$ receptor mRNA and protein have been reported to be relatively low in mouse brain (Hirst *et al.*, 2003).

Studies using light and electron microscopic techniques have shown rat 5-HT$_6$ receptor immunoreactivity to be associated only with neurones (Hamon *et al.*, 1999) and are present in hippocampal cell dendrites (Gerard *et al.*, 1997). Hamon *et al.* (1999) found that receptor immunoreactivity was localized in the strata oriens and radiatum of the CA1 region and the molecular layer of the dentate gyrus oriens. However, the pyramidal and granular cell layers did not show any immunolabeling. Taken together with the presence of 5-HT$_6$ mRNA in the latter areas (Ward *et al.*, 1995), this finding indicated that 5-HT$_6$ receptors may be transported from the pyramidal and granule cell bodies to dendritic areas.

Ablation of serotonergic neurones in rat brain with 5,7-dihydroxytryptamine did not affect levels of 5-HT$_6$ receptor mRNA in the hippocampus, striatum and nucleus accumbens (Gerard *et al.*, 1996), suggesting that it is not present as an autoreceptor on serotonergic nerve terminals. Lesioning of the nigro-striatal pathway with 6-hydroxydopamine also did not alter the level of the 5-HT$_6$ receptor antagonist, [^{125}I]-SB-258585, binding in any of the brain regions examined, indicating that 5-HT$_6$ receptors are not located on dopaminergic neurones (Roberts *et al.*, 2002). Similarly, lesioning of the cholinergic system with the immunotoxin, 192-IgG-saporin, did not alter the density of 5-HT$_6$ receptor mRNA or protein, arguing against 5-HT$_6$ receptor localization on cholinergic neurones (Marcos *et al.*, 2006). Neurones that express dendritic 5-HT$_6$ receptors have been shown to innervate glutamic acid decarboxylase-positive cells. Using a dual-labeling immunohistochemical technique, 5-HT$_6$ receptors were colocalized with GABAergic neurones in rat hippocampus in more than 20% of 5-HT$_6$ immunoreactive neurones (Fone *et al.*, 2002).

When compared with other members of the serotonin receptor family, the 5-HT$_6$ receptor shows the most sequence homology with 5-HT$_2$ receptors (Setola and Roth, 2003). Human and rat 5-HT$_6$ receptors have an 89% amino acid sequence homology (Kohen *et al.*, 1996). The mouse amino acid sequence shows 97% homology with that of the rat and 89% similarity to the human sequence (Hirst *et al.*, 2003). Despite this high degree of receptor homology between rats, humans and mice, important differences exist between the mouse 5-HT$_6$ receptor subtype and rat and human homologs of this receptor (Hirst *et al.*, 2003). Thus, the 5-HT$_6$ receptor is not expressed at high levels in the basal ganglia of the mouse, unlike the human or rat, and in this species, it is also expressed at much lower levels in a number of other brain regions (Hirst *et al.*, 2003). Furthermore, site-directed mutagenesis experiments performed by Hirst *et al.* (2003) revealed that the binding pocket where antagonists and agonists bind is quite different in the mouse 5-HT$_6$ receptor compared with those present in the human or rat 5-HT$_6$ receptor subtypes. Together, these differences indicate that while the rat is a good model species to predict the pharmacology of 5-HT$_6$ receptor ligands in humans, much more caution needs to be exercised when evaluating data obtained from mice.

III. Preclinical Characterization of 5-HT$_6$ Receptor Ligands in Obesity and Related Metabolic Disorders

A. Acute Effects on Feeding Behavior

The first indication that 5-HT$_6$ receptors may have a role in regulating food intake came from studies performed by Fone and colleagues (Woolley *et al.*, 2001). These researchers reported that intracerebroventricular administration of 5-HT$_6$ antisense oligonucleotides to rats decreased their feeding behavior (Bentley *et al.*, 1997; Woolley *et al.*, 2001). These researchers also performed behavioral experiments with some of the first bioavailable and brain-penetrant 5-HT$_6$ receptor ligands, that is, Ro 04-06790 and SB-271046 (Table III). In addition to reporting their cognitive enhancing properties, they also noted that both 5-HT$_6$ receptor antagonists produced dose-dependent reductions of food intake when given acutely to rats (Bentley *et al.*, 1999; Woolley *et al.*, 2004). Further evidence that

Table III
Profiles of 5-HT$_6$ Receptor Ligands that have been Evaluated in Various Rodent Models of Obesity.

Compound	Company	5-HT$_6$ Affinity (Rat)	Functional Activity
Ro 04-6790	Roche	28.8 nM[a,b]	Antagonist
MEM68626	Roche	NPI	Antagonist
MEM68753	Roche	1.0 nM[c]	Antagonist
SB-271046	GSK	0.76 nM[a,b]	Antagonist
SB-742457	GSK	0.25 nM[b]	Antagonist
SUVN503/SUVN 504	Suven	3.2 nM[d]	Antagonist
SUVN51005	Suven	19.2 nM[e]	Antagonist
BVT.5182	Biovitrum	0.25 nM[f]	Antagonist
BVT.71346 (BVT.A)	Biovitrum	1.0 nM[g]	Antagonist
PRX07034	Predix	4.0 nM[h]	Antagonist
E6837	Esteve	4.3 nM[i]	Partial agonist

NPI: No public information.
[a] Hirst *et al.* (2003).
[b] Upton *et al.* (2008).
[c] Callahan *et al.* (2009).
[d] Bhyrapuneni *et al.* (2007).
[e] Shanmuganathan *et al.* (2008).
[f] Svartengren *et al.* (2004).
[g] Svartengren *et al.* (2007).
[h] Gannon *et al.* (2006a, 2006b).
[i] Fisas *et al.* (2006).

the 5-HT_6 receptor is a viable target for antiobesity drug research was provided by the 5-HT_6 receptor knockout mouse. Tecott and Brennan (2000), published a US patent describing this genetic phenotype, and in collaboration with scientists from Biovitrum, showed that the 5-HT_6 receptor knockout mouse was resistant to dietary-induced obesity when maintained on a high-fat diet (Caldirola, 2003). More recently, Frassetto *et al.* (2008) developed a mutant strain of mouse with a nonfunctional 5-HT_6 receptor. They showed that this phenotype maintained a normal weight and was healthy when fed standard laboratory chow, but when it was given access to a high-fat diet, the 5-HT_6 receptor mutant mouse consumed 8% less food and weighed 35% less than the wild-type controls. In the intervening period since Fone and colleagues investigated Ro 04-06790 and SB-271046, many 5-HT_6 receptor ligands have been studied for their effects on feeding behavior in normal and obese rats, but much of this information is not in the public domain. The list of 5-HT_6 receptor ligands that have been tested in animal models of feeding behavior and obesity for which published information is available is provided in Table III.

We have characterized the metabolic effects of a wide range of 5-HT_6 receptor ligands in our laboratory with functional activities ranging from antagonism through to high-efficacy partial agonism (>50% intrinsic efficacy). From the evidence that is presented, it will be clear that both antagonists and agonists of the 5-HT_6 receptor show considerable promise as novel antiobesity drug candidates, but the reason for this pharmacological enigma has not yet to been satisfactorily explained; a detailed discussion of this point is provided in the review by Heal *et al.* (2008).

For weight loss to occur in humans or animals, they must exist in a state of "negative energy balance," that is, the number of calories expended must exceed the number of calories consumed. This state can be achieved by a reduction in food intake (calories consumed), an increase in energy output (calories expended), or an additive combination of both effects. When studying the actions of drugs and test compounds on food intake in rodents, it is important to employ naturalistic experimental conditions. Rodents are nocturnal feeders and consume ~60% of their daily intake during the dark phase. Therefore, it is of little relevance to evaluate the effects of drugs on food intake in the light phase when rodents sleep with only occasional bouts of feeding. To determine the physiologically relevant effects of drugs on this behavior, the animals need to be kept on a reverse-phase lighting regime with free access to food. The compounds need to be administered just into the dark phase when the animals are active and consume most of their daily food intake. In this paradigm, the 5-HT_6 receptor antagonist, PRX07034, produced a dose-dependent reduction in rats' food intake when administered acutely, and at the highest dose tested (30 mg/kg i.p.), the compound reduced the cumulative food of the rats over a full 24 h period (Fig. 2). Providing that energy expenditure is unaltered, this result would indicate that the compound is able to

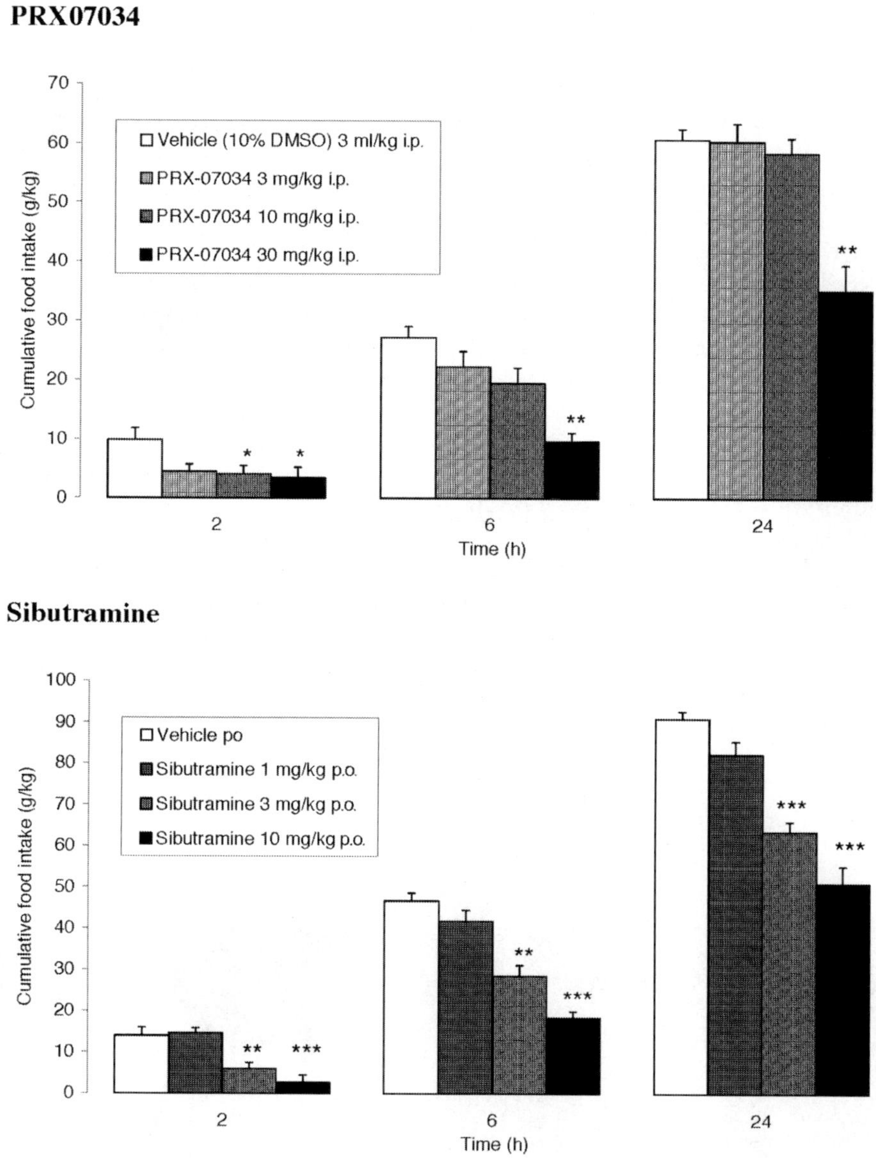

FIG. 2. Effect of acute administration of the 5-HT$_6$ receptor antagonist, PRX07034, and the reference antiobesity drug, sibutramine, on the cumulative food intake of normal rats. Results are expressed as mean + SEM. PRX07034 experiment – rats were intraperitoneally injected twice with PRX07034 or vehicle at 0 and 6 h. Sibutramine experiment – rats were orally administered sibutramine or vehicle at 0 h. Significant differences from the vehicle-treated control are denoted by $^{*}p < 0.05$, $^{**}p < 0.01$, and $^{***}p < 0.001$ (one-way ANOVA by Dunnetts' multiple comparison test). Data taken from Gannon *et al.* (2006a, 2006b) (PRX07034) and Jackson *et al.* (1997).

induce a state of negative energy balance, which, if maintained when the drug is given repeatedly, would result in a decrease of body weight. For comparative purposes, the hypophagic action of PRX07034 is shown next to that of the clinically proven antiobesity drug, sibutramine. It is evident that the overall efficacy and potency of the 5-HT$_6$ receptor antagonist are considerably less than those of this serotonin and noradrenaline reuptake inhibitor (SNRI), particularly at the 24 h time-point (Fig. 2). Two tentative conclusions can be drawn from this observation. First, to achieve optimal efficacy in repeat-dose studies, it may be better to administer PRX07034 twice daily, and second, the pattern of effects of 5-HT$_6$ receptor ligands on feeding behavior may be very different from those of the SNRIs. These findings are consistent with those of other groups who have shown that a wide range of 5-HT$_6$ receptor antagonists reduce food intake in normal rats when administered acutely, including BVT.5182 (Bokare *et al.*, 2008; Svartengren *et al.*, 2004), PRX07034, SB-271046, SB-399885, SB-742457 (Bokare *et al.*, 2008), SUVN504 (Bokare *et al.*, 2008; Sastry *et al.*, 2007), SUVN51005 (Shanmuganathan *et al.*, 2008), and BVT.71346 (Svartengren *et al.*, 2007).

As shown in Fig. 3, drugs can suppress feeding behavior through a wide variety of mechanisms, but the only ones that are clinically acceptable for an antiobesity drug candidate are to reduce hunger and/or enhance satiety. Further experiments revealed that nonspecific suppression of feeding by 5-HT$_6$ receptor ligands as a result of motor side effects can be discounted. Thus, Svartengren *et al.* (2004) reported that the 5-HT$_6$ receptor antagonist, BVT.5182, dose dependently

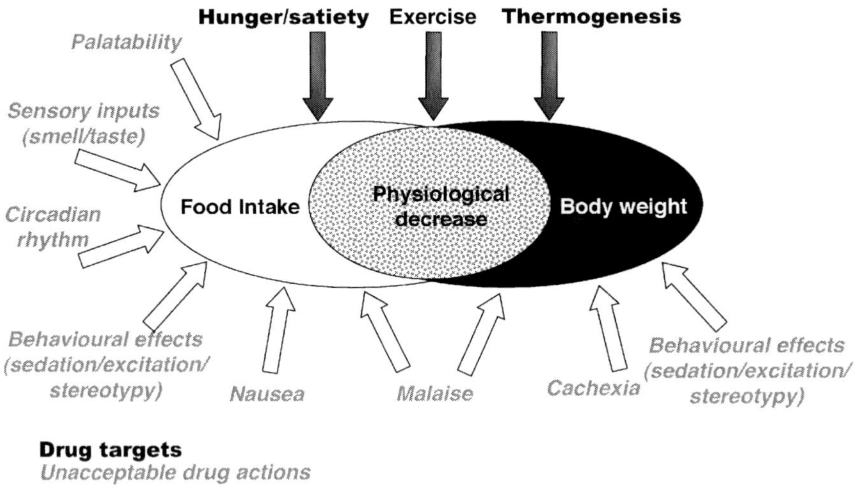

FIG. 3. Factors controlling food intake and body weight. Hunger/satiety and thermogenesis are the only physiological mechanisms suitable for the development of new antiobesity drugs.

decreased the food intake of normal rats and obese, ob/ob mice without inducing sedation, behavioral activation, or stereotypy. Similarly, Fisas *et al.* (2006) reported that the 5-HT$_6$ receptor partial agonist, E6837 (30 mg/kg p.o.), produced a marked reduction of feeding in both normal and obese rats with no effect on locomotor activity.

Rodents lack an emetic reflex and consume inert materials, for example, bedding, to expel noxious substances from the gastrointestinal tract. This defensive strategy can be studied by allowing the animals access to kaolin. If a compound induces nausea or malaise, the animals will consume kaolin in an attempt to eliminate the compound (pica behavior). Svartengren *et al.* (2004) reported the effects of BVT.5182 on food intake and kaolin ingestion in normal rats. As shown in Fig. 4, BVT.5182 produced a dose-dependent reduction in food intake at doses between 3 and 30 mg/kg s.c. The concomitant measurement of kaolin ingestion revealed that the two lower doses of BVT.5182 did not induce kaolin consumption, demonstrating that their actions to reduce food intake were not mediated through gastrointestinal disturbance. The observation that BVT.5182 initiated kaolin consumption at the highest dose clearly delineates between the clinically acceptable and unacceptable actions of the compound on feeding. Fisas *et al.* (2006) reported that E6837 did not induce kaolin consumption in normal rats at doses $\leq$60 mg/kg p.o. Complementary to these findings, Gannon *et al.* (2006a) reported that PRX07034 (10 mg/kg) did not reduce the consumption by normal rats of either saccharine or sodium ingestion after moderate sodium depletion, demonstrating that its hypophagic effect was also not due to conditioned taste aversion.

Studies with 5-HT$_6$ receptor ligands indicate that they reduce food intake by enhancing satiety (meal termination) rather than reducing appetite (meal initiation). In an analysis of meal patterns, BVT.5182 (5 and 15 mg/kg s.c.) had no effect on the latency to eat (a measure of appetite), but it reduced the number of meals consumed and increased the time interval between them indicating both enhanced satiety and prolonged maintenance of satiation. The "satiety sequence" is the name given to the characteristic pattern of behaviors that rodents display when feeding, that is, feeding followed by grooming and ultimately postprandial resting (Antin *et al.*, 1975). In a time-sampled variant of this model, PRX07034 (10 and 30 mg/kg i.p.) produced a pattern of changes that was consistent with the profile of a drug that reduces food consumption by enhancing satiety; that is, it reduced the time spent eating, there was a more rapid termination of feeding, and it increased postprandial resting Gannon *et al.* (2006a).

Overall from this evaluation, it can be concluded that 5-HT$_6$ receptor ligands reduce food intake, but more importantly, the compound-induced hypophagia is a specific effect that is mediated by enhancing the natural physiological process of behavioral satiety.

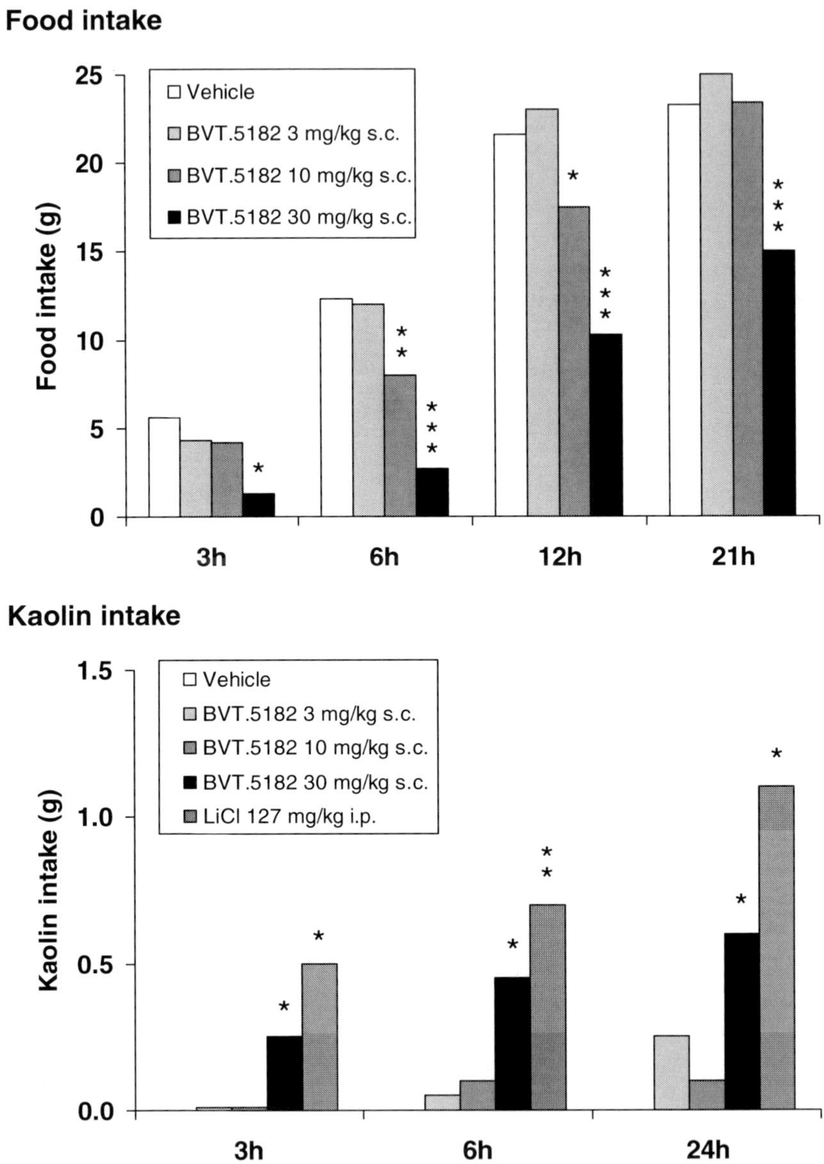

FIG. 4. Effect of acute administration of various doses of the 5-HT$_6$ receptor antagonist, BVT.5182, on food consumption and kaolin ingestion in normal rats. Food intake data recorded by computerized rodent feeding meter. Test compounds administered at dark onset. Rats were adapted to kaolin for 10 days before administration of drug. *$p < 0.05$, **$p < 0.01$, ***$p < 0.001$ versus vehicle. Data adapted from poster by Svartengren *et al.* (2004).

B. Effects in Rodent Models of Human Obesity

Rodents are particularly well suited for the study of human obesity. They are omnivorous and will maintain a reasonably healthy weight and body composition during adolescence and early adulthood when maintained on a nutritionally well-balanced diet. Like humans, rodents find unhealthy, calorie-dense, sweet and/or high fat foods irresistible. Given free access to these food sources, rodents will overconsume and become obese; given restricted, intermittent access to them, they will binge-eat (Berner *et al.*, 2008; Boggiano *et al.*, 2007; Corwin, 2004, 2006; Wojnicki *et al.*, 2008). Female rats are particularly appropriate for studying obesity in maturity because unlike the males that continue to grow throughout their life cycle, females reach and maintain a reasonably stable adult body weight. We have developed a rat model of human obesity by switching mature, weight-stable, female rats from laboratory chow to a "junk food" diet consisting of high-fat chow, ground milk chocolate, and ground roasted, salted peanuts (Dickinson *et al.*, 2001; Heal and Jagger, 2005). These food sources are made available to the rats *ad libitum* and provided in individual food pots to allow a cafeteria-style selection and consumption by the animals. As shown in Table IV, the syndrome of obesity and associated comorbidities in the diet-induced obese (DIO), female rat is identical to human obesity with insulin resistance (face validity), and the factors

Table IV
Face and construct validity of the diet-induced obese (DIO) female rat model of human obesity with insulin resistance.

Characteristics	DIO rat	Human obesity
Sex	Female	Obesity occurs equally in both sexes, but females predominate in seeking drug treatment
Mature–onset, gradually increasing obesity	✓	✓
Polygenic basis	✓	✓
Calorie-dense, high fat, "junk food" diet	✓	✓
Ad libitum access to different, highly palatable food sources.	✓	✓
Marked visceral adiposity	✓	✓
Leptin resistant	✓	✓
Insulin resistant	✓	✓
Hyperglycaemic	✗	✗/✓
Hypertensive	✓	✓

✓ = Present; ✗ = Absent; ✗/✓ = Type 2 diabetes may or may not be comorbid with obesity.

contributing to their initiation and maintenance of obesity are similar in both species (construct validity). "Predictive validity" is the extent to which outcomes observed in animal models translate accurately to the human situation. In anti-obesity drug research, the most important aspect of predictive validity is whether the pharmacological effects of compounds on body weight and cardiometabolic risk factors in the animal model reliably translate into therapeutic benefits in clinical trials. The DIO female rat model is exceptionally advantaged in this respect because it possesses not only qualitative but also quantitative, predictive validity. Thus, a wide range of antiobesity drugs and drug candidates, which have been evaluated in the clinic, have been studied in the DIO female rat. Every compound that has been reported to decrease body weight in human subjects has also been shown to induce weight loss in this obese rat model (Fig. 5). Furthermore, as shown in Fig. 5, there is a high degree of correlation ($r^2 = 0.95$)

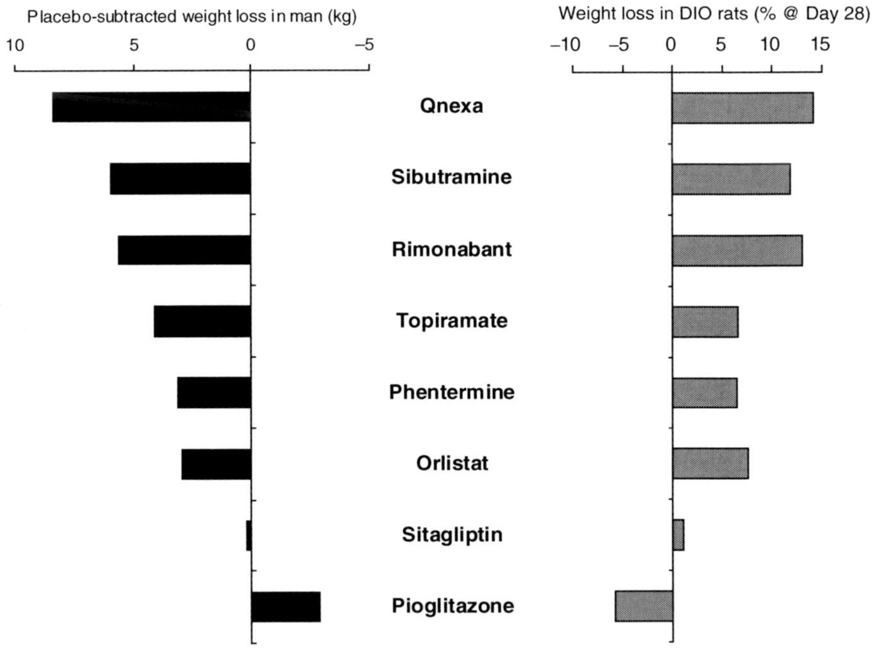

FIG. 5. Relative weight-loss efficacy in humans of various antiobesity drugs – predictive validity of the DIO female rat model of human obesity. The panel on the left shows the reported weight changes observed in clinical trials and meta-analyses of various metabolic drugs, that is, the antiobesity drugs, Qnexa, sibutramine, rimonabant, topiramate, phentermine, and orlistat, and the antidiabetic drugs, sitagliptin and pioglitazone (see text for source references). The panel on the right shows the changes in body weight induced by the same drugs in the diet-induced obesity (DIO) female rat model of human obesity and insulin resistance (data reported in Fig. 9 or on file at RenaSci Consultancy). The very high degree of correlation between these two variables is shown above. Correlation $r^2 = 0.95$.

between the magnitude of weight loss that these antiobesity drugs produced in clinical trials and their rank order for inducing weight loss in the DIO female rat. The predictive validity of the model also extends into drug-induced side effects. Sitagliptin (a dipeptidyl peptidase-4 [DPP-4] inhibitor) and pioglitazone (a thiazolidinedione) are examples of two of the major classes of antidiabetic drugs. Although DPP-4 inhibitors and thiazolidinediones improve insulin sensitivity in Type 2 diabetes, the former are weight-neutral, while the latter induce a degree of weight gain (Foley and Jordan, 2010; Shah and Mudaliar, 2010). Consistent with the clinical findings (Aronoff *et al.*, 2000; Aschner *et al.*, 2006; Goldstein *et al.*, 2007; Nissen *et al.*, 2008), although sitagliptin and pioglitazone both reduced insulin resistance in the DIO female rat (data on file, RenaSci), sitagliptin produced an improvement in glycemic control without altering body weight. Pioglitazone only achieved this outcome with significant weight gain. In addition to predicting weight-loss efficacy, many of the other therapeutic benefits of metabolic drugs in obesity are also observed in the DIO female rat, for example, reductions in plasma lipids, decreased insulin resistance, an increased ability to handle glucose loads, reductions in plasma leptin and other biomarkers (see later sections).

It is clear from this brief summary that the DIO female rat is a valuable model for evaluating novel antiobesity drug candidates, and a high degree of confidence can be placed on predictions that therapeutic benefits observed in this rat model will translate both qualitatively and possibly quantitatively into the clinical setting.

In terms of the evaluation of 5-HT$_6$ receptor ligands, the effects of prolonged administration of both antagonists and partial agonists have been investigated in the DIO female rat. The effects on body weight of treatment with PRX07034 (5-HT$_6$ receptor antagonist), E6837 (5-HT$_6$ receptor partial agonist), and sibutramine (SNRI) are compared in Fig. 6. Both 5-HT$_6$ receptor drug candidates produced gradual decreases in body weight with trajectories that indicated greater weight reductions were potentially possible if treatment had been continued. The weight reductions produced by PRX07034 and E6837 were greater than that evoked by the comparator antiobesity drug, sibutramine. Furthermore, the pattern of weight change obtained with sibutramine was very different from those observed with the 5-HT$_6$ receptor ligands. The latter produced a delayed, but almost linear, fall in body weight, whereas sibutramine caused a rapid fall in weight followed by a sustained period of weight-loss maintenance relative to placebo. These different patterns of body weight effects are entirely explained by the food intake results (Fig. 7). Sibutramine markedly suppressed the food intake of DIO female rats at the start of dosing, but gradually this effect diminished and feeding returned to control levels after 10–12 days; this explains the rapid fall in body weight followed by a sustained period of weight-loss maintenance. In contrast, food intake was reduced to a lesser extent by the 5-HT$_6$ receptor ligands, and in the case of E6837, the delay in reaching its nadir is clearly evident. However, the

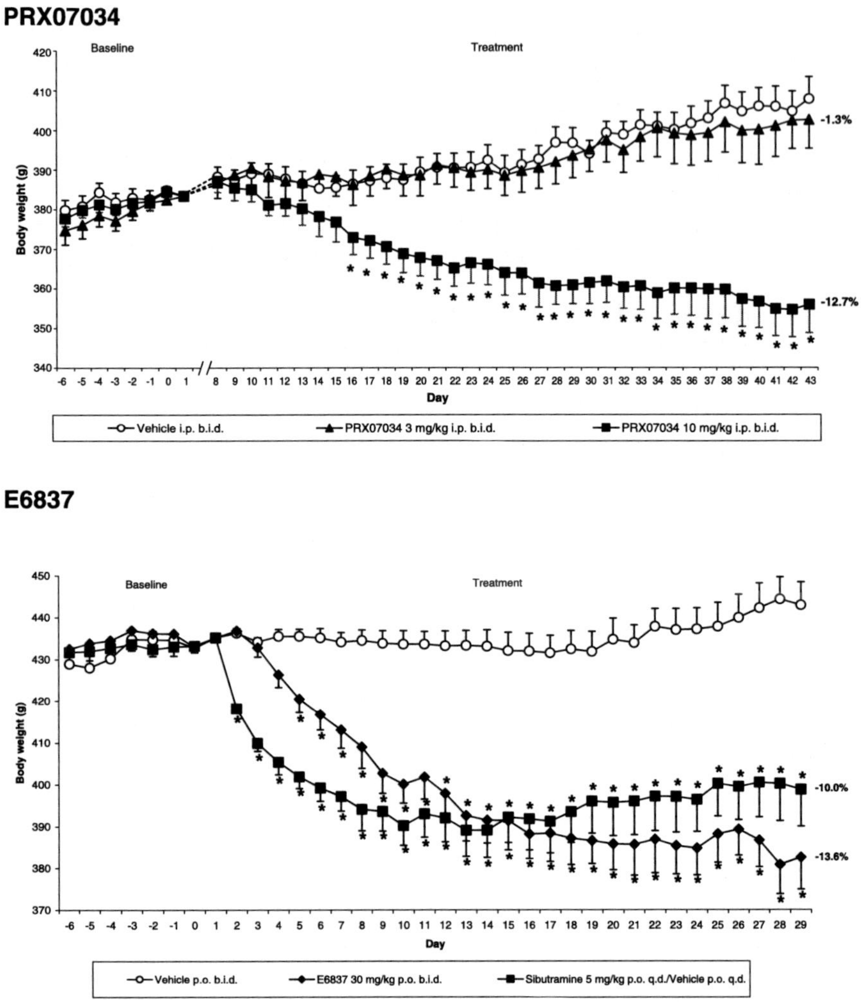

FIG. 6. Effects of the 5-HT_6 receptor antagonist, PRX07034, and partial agonist, E6837, on body weight in obese, female rats maintained on a simplified cafeteria diet. Data are adjusted means. SEMs are calculated from residuals of the statistical model. Drug treatment commenced on Day 1. PRX07034 was dosed at 10 and 30 mg/kg p.o. b.i.d. on Days 1–7 (data not shown) followed by 3 and 10 mg/kg i.p. b.i.d. on Days 8–42. Data analyzed by ANCOVA with Day 1 as covariate. Significant difference from vehicle assessed using Dunnett's test (PRX07034 and E6837), or the multiple *t*-test (sibutramine). Significant differences from the control group are denoted by $*p < 0.05$. Data taken from Gannon *et al.* (2006a,b) (PRX07034) and Fisas *et al.* (2006) (E6837 and sibutramine).

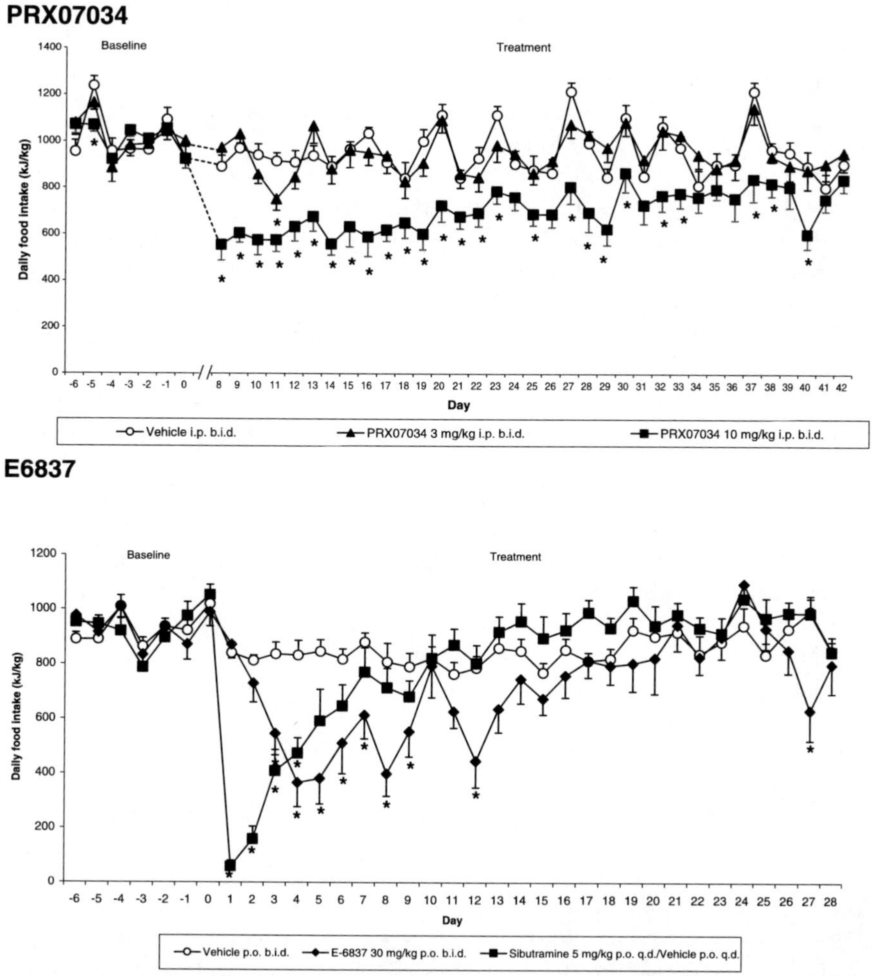

FIG. 7. Effects of the 5-HT$_6$ receptor antagonist, PRX07034 and the 5-HT$_6$ partial agonist, E6837, on food intake in obese, female rats maintained on a simplified cafeteria diet. Data are adjusted means. SEMs are calculated from residuals of the statistical model. Drug treatment commenced on Day 1. PRX07034 was dosed at 10 and 30 mg/kg p.o. b.i.d. on Days 1–7 (data not shown) followed by 3 and 10 mg/kg i.p. b.i.d. on Days 8–42. Data analyzed by ANCOVA with baseline food intake as covariate. Significant difference from vehicle assessed using Dunnett's test (PRX07034 and E6837), or the multiple *t*-test (sibutramine). Significant differences from the control group are denoted by $*p < 0.05$. Data taken from Gannon *et al.* (2006a) (PRX07034) and Fisas *et al.* (2006) (E-6837 and sibutramine).

decreases induced by either PRX07034 or E6837 were maintained for a much longer period with significant reductions induced by PRX07034 present almost to the end of the dosing phase (42 days). These profiles of reduced food intake explain why weight loss evoked by 5-HT_6 receptor ligands is not only gradual in onset, but also follows a linear, downward trajectory.

The DIO female rat model of human obesity has the disadvantage that middle-aged female rats need to be maintained on the simplified cafeteria diet for at least 12 weeks before their obesity stabilizes. In addition to this technical limitation, it has to be acknowledged that obesity is not a clinical condition that is restricted either to females or to middle-age. To address these issues, we have developed a model of obesity in young adult, male rats maintained on a high-fat diet. This obesity model does not require prolonged exposure to the calorie-dense diet before compound testing, and it mimics human obesity occurring in adolescence or early adulthood when excess adiposity is occurring at the same time as growth. The 5-HT_6 receptor antagonists, PRX07034 and MEM68626, have both been evaluated in the high-fat-fed, obese, male rat as shown in Fig. 8. Dose-dependent reductions in body weight were produced by PRX07034 (30 and 100 mg/kg p.o. b.i.d.) and MEM68626 (10–100 mg/kg p.o. b.i.d.) with decreases of 11.8 and 11.6%, respectively, observed after 28 days of treatment at the highest doses of these compounds (Fig. 8). In this model, the reference comparator antiobesity drugs, sibutramine, and the CB_1 receptor antagonist, rimonabant, decreased body weight ~10.4 and 6.5%, respectively, demonstrating that the clinical efficacy of these 5-HT_6 receptor antagonists is predicted to be as good as, if not better than, existing antiobesity drugs. The profiles of reduced food intake produced by these drugs in high-fat-fed male rats followed patterns similar to those observed in the DIO female rat (Fig. 7; Gannon *et al.*, 2006a; Murray *et al.*, 2008; Shacham *et al.*, 2006).

Various other 5-HT_6 receptor ligands have been evaluated in rat models of human obesity. We have also observed that SB-742457 (10 mg/kg i.p. b.i.d.) produced a very large 9.8% reduction in body weight in the high-fat-fed obese male rat after only 7 days of treatment (Sargent and Moore, 2009). Biovitrum's 5-HT_6 receptor antagonists, BVT.5182 and BVT.71346, have also been shown to produce dose-dependent and substantial weight loss in the high-fat-fed obese rat (Svartengren *et al.*, 2004, 2007). Although the specific model employed was not stated, Shanmuganathan *et al.* (2008) reported that the 5-HT_6 receptor antagonist, SUVN51005, reduced body weight gain in DIO rats.

Since the mouse 5-HT_6 receptor homolog differs significantly from the human and rat 5-HT_6 receptor subtypes, and because the 5-HT_6 receptor has a different distribution in the mouse brain from that in the rat and human (Hirst *et al.*, 2003; Setola and Roth, 2003), the predictive validity of data obtained in mouse models of obesity has been treated with a degree of caution. However, in spite of the potential criticisms, several 5-HT_6 receptor antagonists have been tested in both rats and mice with similar outcomes observed in both species. Thus, the 5-HT_6 receptor

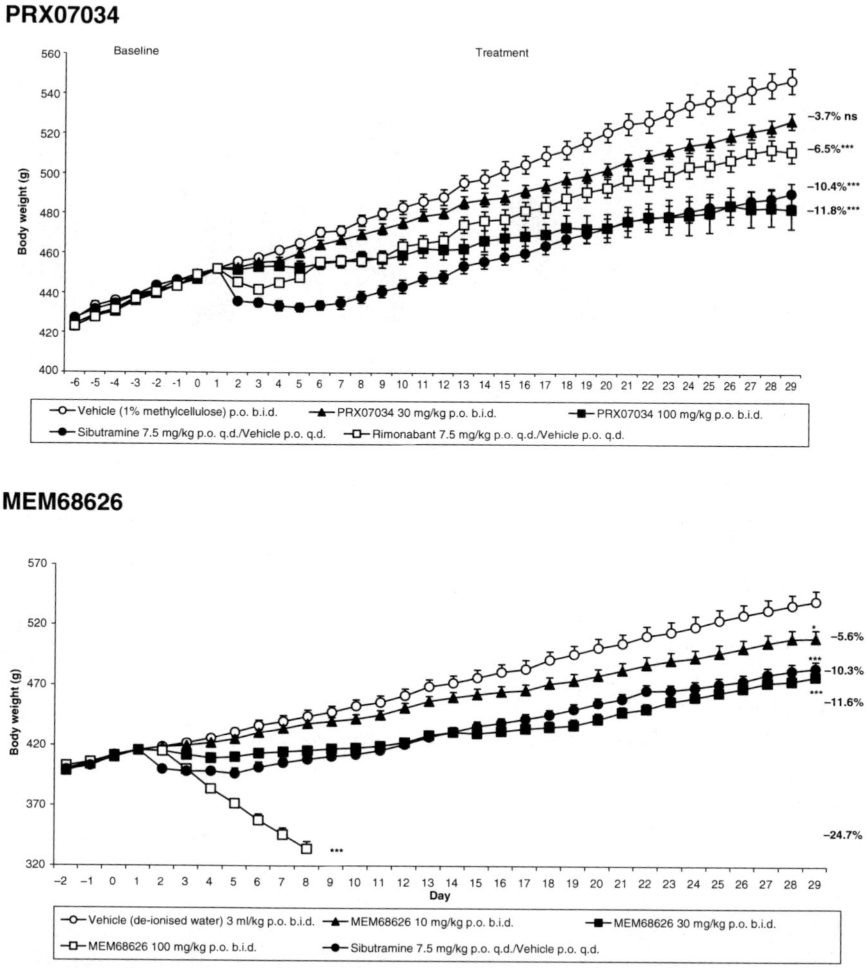

FIG. 8. Effects of the 5-HT$_6$ receptor antagonists, PRX07034 and MEM68626, on body weight in obese, male rats maintained on a high-fat diet. Data are adjusted means. SEMs are calculated from residuals of the statistical model. Drug treatment commenced on Day 1. Data analyzed by ANCOVA with Day 1 as covariate. Significant difference from vehicle on Day 29 assessed using Dunnett's test (PRX07034), Williams' test (MEM68626) or the multiple *t*-test (sibutramine and rimonabant). Significant differences from the relevant control group are denoted by $*p < 0.05$ and $***p < 0.001$. Data taken from Gannon *et al.* (2006b), Shacham *et al.* (2006) (PRX07034) and Murray *et al.* (2008) (MEM68626).

antagonist, BVT.5182, has been shown to decrease the food consumption of *ob/ob* and DIO mice (Svartengren *et al.*, 2004), and in the latter model, to produce substantial decreases in body weight when given repeatedly (Svartengren *et al.*, 2004). The pharmacological effects of BVT.5182 in the mouse are identical to

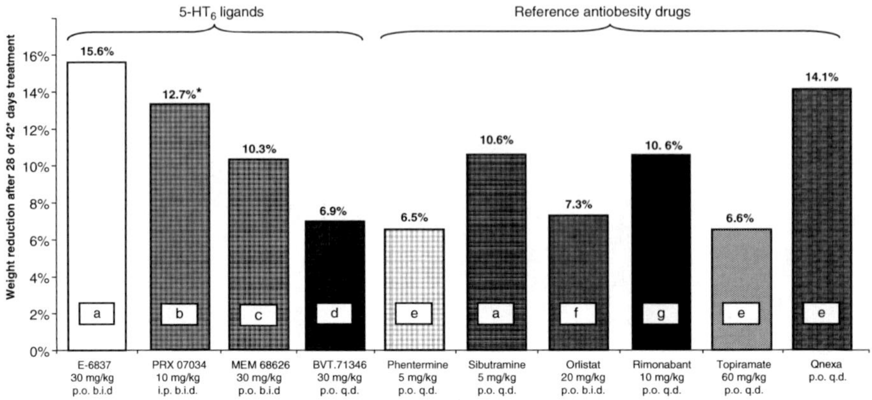

FIG. 9. Weight loss in rat models of human obesity produced by various 5-HT$_6$ receptor ligands in comparison to clinically proven antiobesity drugs. Data taken from (a) Fisas *et al.* (2006) and Murray *et al.* (2008); (b) Gannon *et al.* (2006a,b); (c) Murray *et al.* (2008); (d) Svartengren *et al.* (2007); (e) Jackson *et al.* (2007); (f) Jackson *et al.* (2004); (g) Jackson *et al.* (2005) and Murray *et al.* (2008). *Weight loss at Day 42. (For interpretation of the references to color in this figure legend, the reader is referred to the Web version of this chapter.).

those observed when the compound was administered to normal and DIO rats (Svartengren *et al.*, 2004). Similarly, SUVN504 has been reported to decrease food intake in both rats (Bokare *et al.*, 2008) and mice (Sastry *et al.*, 2007) and to reduce the body weight of diet-induced mice when administered chronically (Sastry *et al.*, 2007).

To summarize, 5-HT$_6$ receptor antagonists and partial agonists reduce food intake in rodent models of obesity and produce reductions in body weight in these models that are at least as great as those evoked by a number of clinically proven antiobesity drugs. Figure 9 provides an illustrative comparison of the weight losses produced by various 5-HT$_6$ receptor ligands tested in the DIO female rat and the high-fat-fed male rat models of human obesity in comparison to the effects that have been observed with a range of reference comparator, antiobesity drugs and drug candidates.

C. Effects on Cardiometabolic Risk Factors

One of the first steps when observing drug-induced weight loss in animals is to define which components of body composition account for this reduction. A decrease in adiposity is a clinically beneficial outcome, whereas a reduction mediated through a loss of protein (muscle wasting or cachexia) or water (dehydration) is not only nonbeneficial but also harmful. Body composition analysis performed using a range of techniques has shown that weight loss in obese rodents produced by 5-HT$_6$ receptor antagonists and the partial agonist, E6837, is mediated by a selective reduction of body fat with no loss of protein (Fisas *et al.*, 2006; Gannon *et al.*, 2006b; Shacham *et al.*, 2006; Svartengren *et al.*, 2007).

It is well known that fat depots in the abdomen (visceral fat) carry a much greater burden of cardiovascular risk than subcutaneous fat (Nielsen and Jensen, 1997; Peiris *et al.*, 1989). In addition to demonstrating that 5-HT$_6$ receptor ligands produce weight loss in obesity through a selective reduction of fat mass, as shown in Fig. 10, part of the effect of the 5-HT$_6$ receptor antagonist, PRX07034, was attributable to a loss of visceral fat. Reductions in visceral fat mass in obese rats and mice have also been reported after treatment with BVT.5182 (Svartengren *et al.*, 2004) and MEM68626 (Murray *et al.*, 2008).

A drug-induced reduction of body weight is merely a cosmetic effect if the weight loss is not accompanied by commensurate improvements in a range of cardiometabolic risk factors. It is for this reason that the guidelines for developing novel antiobesity drugs issued by EMA (2007) and the FDA (2007) stipulate that reductions in visceral adiposity, blood pressure, increased glycemic control and improvements in plasma lipid profiles are critical clinical outcomes when evaluating the medical benefits of new drugs for weight management. Many, but not all, of these cardiometabolic disturbances are also present in rodent models of human

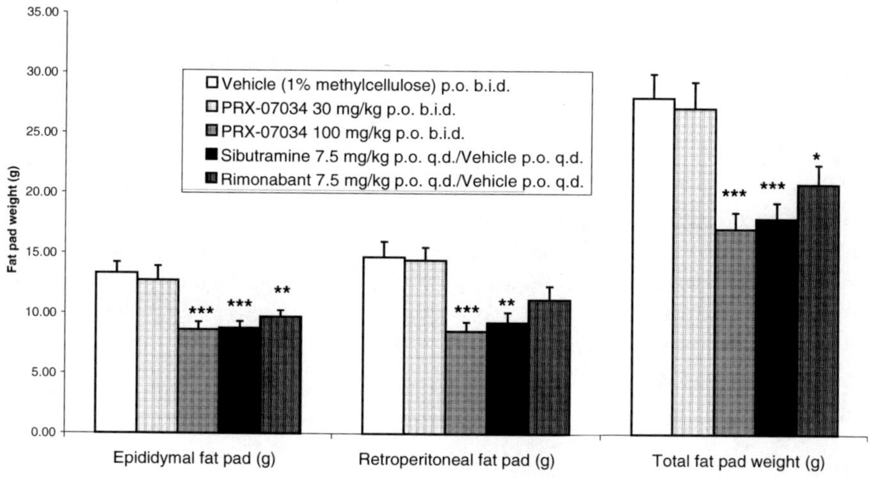

FIG. 10. Effect of chronic administration of the 5-HT$_6$ receptor antagonist, PRX07034, or the reference antiobesity drugs, sibutramine and rimonabant, on fat pad weight in obese, male rats maintained on a high-fat diet. Data are simple means + SEMs, n = 9–10 rats/group. Drug doses are expressed as the free base. Significant difference from the vehicle-treated control are denoted by $^{*}p < 0.05$, $^{**}p < 0.01$, and $^{***}p < 0.001$ (one-way ANOVA followed by Dunnetts' multiple comparison test). Data taken from Gannon *et al.* (2006b) and Shacham *et al.* (2006).

obesity (Table IV) and will improve significantly when the excess adiposity of the rats is reduced, for example, reductions in visceral adiposity and blood pressure, increased glycemic control and improvements in plasma lipid profiles. Exceptions are alterations in the plasma concentrations of LDL-C ("bad cholesterol") and HDL-C ("good cholesterol") because this aspect of lipid metabolism is very different between humans and rodents (Bergen and Mersmann, 2005). The cardiometabolic benefits of weight loss induced by treatment with a range of 5-HT$_6$ receptor ligands have been studied in various rodent models of obesity and the published findings are reported in Table V. There is good agreement across the studies that 5-HT$_6$ receptor drug candidates decrease overall adiposity and reduce the visceral fat depots that carry a high degree of cardiovascular risk. The reduction in white adipocyte mass is reflected by consistent decreases in plasma levels of the secreted hormone, leptin, although it is uncertain whether this effect translates into decreased leptin resistance. In the rodent models of obesity coupled with insulin resistance, plasma insulin concentrations are generally moderately elevated, but the animals are rarely hyperglycemic or diabetic. Improvements in glycemic control produced in obese rodents by 5-HT$_6$ receptor ligands are generally demonstrated by reductions in the concentration of fasting plasma insulin with no change in fasting plasma glucose (Table V). In the case of E6837, its ability to decrease insulin resistance has been confirmed by results from an oral glucose tolerance test (Fisas *et al.*, 2006). We have not observed any benefit of the weight loss evoked by 5-HT$_6$ receptor ligands on plasma lipids that are relevant in both humans and rodents, that is, free fatty acids, triglycerides, and cholesterol. However, the researchers from Suven have reported that SUVN504 and SUVN51005 decreased plasma triglyceride concentrations (Sastry *et al.*, 2007; Shanmuganathan *et al.*, 2008), although it is unclear in which rodent models of human obesity the compounds were tested.

Together, these results demonstrate that weight loss evoked by 5-HT$_6$ receptor ligands is accompanied by improvements in cardiovascular risk factors. There is no evidence from the data to indicate that these compounds are having any direct beneficial effect on cardiometabolic function that is independent of weight loss. It can, therefore, be concluded that the clinical benefits of the 5-HT$_6$ receptor ligands will be directly correlated with the degree of weight loss that they can produce in the clinical setting.

IV. Closing Remarks and Future Directions

Evidence from a diverse range of sources strongly supports the hypothesis that 5-HT$_6$ receptors have an important role to play in the regulation of food

Table V

WEIGHT-LOSS BENEFITS OF 5-HT$_6$ RECEPTOR LIGANDS ON CARDIOMETABOLIC RISK FACTORS.

Risk Factor	5-HT$_6$ Receptor Drug Candidate BVT.5182	PRX07034	MEM68753	BVT.71346	SUVN504	SUVN51005	E6837
Adiposity	↓[a,b]	↓[a]	↓[a]	↓[a]	ND	ND	↓[a]
Visceral adiposity	↓[a,b]	↓[a]	↓[a]				ND
Leptin	↓[a,b]	↓[a]	↓[a]				↓[a]
Plasma lipids	FFAs ±[b] TGs ±[b]	ND	FFAs ±[a] TGs ±[a] Cholesterol ±[a] Glycerol ±[a]		TGs ↓[b]	TGs ↓[a]	FFAs ±[a] Cholesterol ±[a] Glycerol ±[a]
Insulin resistance	Insulin ↓/±[b] Glucose ±[b]	Insulin ↓[a] Glucose ↓[a]	Insulin ↓[a] Glucose ↓[a]				Insulin ↓/±[a] Glucose ±[a] OGTT ↑[a]
BVT.5182	Data take from Svartengren *et al.* (2003, 2004)						
PRX07034	Data taken from Gannon *et al.* (2006a, 2006b) and Shacham *et al.* (2006)						
MEM68753	Data taken from Callahan *et al.* (2009)						
BVT.71346	Data taken from Svartengren *et al.* (2007)						
SUVN504	Data taken from Sastry *et al.* (2007)						
SUVN51005	Data taken from Shanmuganathan *et al.* (2008)						
E6387	Data taken from Fisas *et al.* (2006)						

↓: Decrease; ↓/±: strong trend for a decrease; ±: no change; ↑: increased sensitivity; ND: not determined; FFAs: free fatty acids; TGs: triglycerides; OGTT: oral glucose tolerance test.[a] Experiments performed in obese rats.

[b] Experiments performed in obese mice.

intake. Studies using a range of 5-HT$_6$ receptor ligands, mostly antagonists, but including the partial agonist, E6837, have revealed that these compounds reduce food consumption in rodents when given acutely and chronically. In rodent models of human obesity, these 5-HT$_6$ receptor compounds reduced body weight and yielded reductions in visceral adiposity and insulin resistance that were consistent with the observed falls in body weight. No adverse effects of drug treatment have been reported. The weight reductions induced by the 5-HT$_6$ receptor ligands in obese rodents are at least as great as those produced by clinically proven antiobesity agents, indicating that the former have the potential to be therapeutically beneficial in the treatment of obesity. According to current information, there are no 5-HT$_6$ receptor ligands in clinical development in the therapeutic indication of obesity at this time, although several of these compounds are being evaluated for the treatment of cognitive dysfunction in Alzheimer's disease and schizophrenia.

Early clinical proof of principle has been obtained for the weight-loss effect of 5-HT$_6$ receptor antagonists in obesity. Epix, which now no longer exists, evaluated the safety and efficacy of PRX07034 in a randomized, double-blind, placebo-controlled, Phase 1b, multiple ascending dose, clinical trial in healthy obese adults (average weight 100 kg or 220 lb). Although the company press release of the findings is no longer available, PRX07034 given at a dose of 600 mg/day for 28 days was reported to produce a placebo-subtracted reduction in body weight of 4–6 lb in these subjects. Although a 4–6 lb body weight reduction might appear to be small or even trivial in subjects weighing over 200 lb, the preclinical experiments clearly indicate that decreases in food intake and body weight evoked by 5-HT$_6$ receptor ligands such as PRX07034 are delayed and gradual, but are maintained during much of the dosing phase. In the DIO female rat, each day of drug administration is approximately equivalent to 1 week in human terms. Thus, sibutramine actively reduces body weight for 12–15 days before the effect plateaus (Fisas *et al.*, 2006; Jackson *et al.*, 2005; Jones *et al.*, 2005; Fig. 6). In clinical trials, active weight reduction on sibutramine treatment takes place over the first 16–20 weeks of treatment (Apfelbaum *et al.*, 1999; Bray *et al.*, 1999; James *et al.*, 2000; Wadden *et al.*, 2005). If the results for the 5-HT$_6$ receptor ligands in the rat models of human obesity translate into man, they indicate that these compounds will reduce food consumption for periods of between 7 months and 1 year with active weight loss over the same period. Thus, although the weight-loss efficacy of the 5-HT$_6$ receptor compounds is likely to be modest at the initiation of treatment, the preclinical data predict that active (i.e. that active weight loss) weight loss will be maintained longer than with existing agents, and ultimately, is likely to be greater. Thus, the 5-HT$_6$ receptor compounds "tick all of the boxes" in terms of proof of concept, weight-loss efficacy, duration of action, and benefits on obesity-related comorbidities.

What is the future for 5-HT$_6$ receptor ligands as novel treatments for the management of obesity? Currently, the only antiobesity drug candidates in late-stage development have been taken this far by small, independent, biotech companies. Most of the major pharmaceutical companies are not actively engaged in R&D on antiobesity drugs. There are several probable reasons for this state of affairs. The "big pharma" players appear to have been discouraged by the conservative attitudes of the regulatory agencies and their advisors toward the approval of new antiobesity drugs. In addition, there has been a general reluctance on the part of healthcare providers to fund antiobesity drug therapy, coupled with an unwillingness by patients to pay for medicines that they feel do not deliver the degree of weight reduction that they are seeking. Together, these factors have resulted in very modest sales figures for marketed antiobesity drugs. These are the likely reasons why big pharma has looked on while the biotech companies have carried the development risks for new drugs in this therapeutic indication, preferring like Takeda, to secure a licensing deal with a biotech partner (Arena) when the drug candidate (lorcaserin) is in the preregistration phase. Some large pharmaceutical companies, for example, Roche, have 5-HT$_6$ receptor compounds with excellent preclinical data to support clinical development as novel antiobesity drugs, but it is unclear whether they are willing to progress the compounds in this clinical indication. Other 5-HT$_6$ programs reside with much smaller companies, for example, AMRI, Suven, and Esteve, who will require collaborations with major pharmaceutical companies to take their compounds through clinical development and onto the market. To date, no deals have been announced, so it is unclear at present whether any of these compounds will progress into full-scale clinical development in obesity.

In summary, there is compelling evidence to show that 5-HT$_6$ receptors have an important function in regulating food intake and body weight not only in animals but also in man. The results from rodent models of human obesity predict that 5-HT$_6$ receptor compounds will be beneficial in reducing human obesity and ameliorating a range of associated cardiometabolic risk factors. There is an urgent need to increase the number and types of medicines available to physicians to help tackle the rising problem of human obesity. For this reason, it is critical that promising 5-HT$_6$ drug candidates should be clinically evaluated to assess their potential contribution and new drugs to treat the obesity epidemic.

Acknowledgments

The authors wish to acknowledge the contribution by all of the scientists from within RenaSci and our industrial sponsors who have collaborated in producing the experimental findings discussed in this review. In particular, we wish to extend our special thanks to Laboratorios Dr Esteve, Epix Pharmaceuticals, and Memory Pharmaceuticals.

References

Ackard, D.M., Neumark-Sztainer, D., Story, M. and Perry, C. (2003). Overeating among adolescents: prevalence and associations with weight-related characteristics and psychological health. *Pediatrics* **111**(1); 67–74.

Antin, J., Gibbs, J., Holt, J., Young, R.C. and Smith, G.P. (1975). Cholecystokinin elicits the complete behavioural sequence of satiety in rats. *J. Comput. Physiol. Psychol.* **89**(7); 748–760.

Apfelbaum, M., Vague, P., Ziegler, O., Hanotin, C., Thomas, F. and Leutenegger, E. (1999). Long-term maintenance of weight loss after a very-low-calorie diet: a randomized blinded trial of the efficacy and tolerability of sibutramine. *Am. J. Med.* **106**(2); 179–184.

Aronoff, S., Rosenblatt, S., Braithwaite, S., Egan, J.W., Mathisen, A.L. andSchneider, R.L. (2000). Pioglitazone hydrochloride monotherapy improves glycemic control in the treatment of patients with Type 2 diabetes: a 6-month randomized placebo-controlled dose-response study. The Pioglitazone 001 Study Group. *Diabetes Care* **23**(11); 1605–1611.

Aschner, P., Kipnes, M.S., Lunceford, J.K., Sanchez, M., Mickel, C. and Williams-Herman, D.E. Sitagliptin Study 021 Group (2006). Effect of the dipeptidyl peptidase-4 inhibitor sitagliptin as monotherapy on glycemic control in patients with Type 2 diabetes. *Diabetes Care* **29**(12); 2632–2637.

Bays, H. (2010). Phentermine, topiramate and their combination for the treatment of adiposopathy (sick fat) and metabolic disease. *Expert Rev. Cardiovasc. Ther.* **8**(12); 1777–1801.

Beckman, J.A., Creager, M.A. and Libby, P. (2002). Diabetes and atherosclerosis. Epidemiology, pathophysiology, and management. *J. Am. Med. Assoc.* **287**(19); 2570–2581.

Bentley, J.C., Marsden, C. A., Sleight, A. J., and Fone, K. C. F. (1997). 5-HT_6 antisense oligonucleotide i.c.v. affects rat performance in the water maze and feeding. Proceedings of the British Association for Psychopharmacology (BAP) Meeting, Cambridge, 13–17 July, P255.

Bentley, J.C., Marsden, C.A., Sleight, A.J. and Fone, K.C.F. (1999). Effect of the 5-HT_6 antagonist Ro 04-6790 on food consumption in rats trained to a mixed feeding regime. *Br. J. Pharmacol.* **126**, 66P.

Bergen, W.G. and Mersmann, H.J. (2005). Comparative aspects of lipid metabolism: impact on contemporary research and use of animal models. *J. Nutr.* **135**(11); 2499–2502.

Berner, L.A., Avena, N.M. and Hoebel, B.G. (2008). Bingeing, self-restriction, and increased body weight in rats with limited access to a sweet-fat diet. *Obesity* **16**(9); 1996–2002.

Bhole, V., deVera, M., Rahman, M.M., Krisnan, E. and Choi, H. (2010). Epidemiology of gout in women: 52-year follow up of a prospective cohort. *Arthritis Rheum.* **62**(4); 1069–1076.

Bhyrapuneni, G., Kurwattimath, V., Lachanna, S., Muddana, N., Benade, V., Saralaya, R., *et al.* (2007). SUVN-504: a potent 5-HT_6 receptor antagonist – an in vivo brain penetration and brain distribution study. Society for Neurosciences Meeting, Abstract 465.3. Available at www.sfn.org.

Block, J.P., Scribner, R.A. and DeSalvo, K.B. (2004). Fast food, race/ethnicity, and income. A geographical analysis. *Am. J. Prev. Med.* **27**(3); 211–217.

Boggiano, M.M., Artiga, A.I., Pritchett, C.E., Chandler-Laney, P.C., Smith, M.L. and Eldridge, A.J. (2007). High intake of palatable food predicts binge-eating independent of susceptibility to obesity: an animal model of lean vs. obese binge-eating and obesity with and without binge-eating. *Int. J. Obes.* **31**(9); 1357–1367.

Bokare, A. M., Shanmuganathan, D., Jayarajan, P., Abraham, P. R., Marshal, V. R., and Nirogi, R. (2008). 5-HT_6 antagonists in animal models of feeding behaviour. Society for Neurosciences Meeting, Abstract 584.8. Available at www.sfn.org.

Bowman, S.A., Gortmaker, S.L., Ebbeling, C.B., Pereira, M.A. and Ludwig, D.S. (2004). Effects of fast-food consumption on energy intake and diet quality among children in a national household survey. *Pediatrics* **113**(1 Pt. 1); 112–118.

Bray, G.A., Blackburn, G.L., Ferguson, J.M., Greenway, F.L., Jain, A.K. and Mendel, C.M., *et al.* (1999). Sibutramine produces dose-related weight loss. *Obes. Res.* **7**(2); 189–198.

Brown, M., Byatt, T., Marsh, T., and McPherson, K. (2010). A prediction of obesity trends for adults and their associated diseases. Analysis from the Health Survey for England 1993–2007. Report by the National Heart Forum. Available at www.heartforum.org.uk/resources/nhf-publications.

Caldirola, P. (2003). 5-HT_6 receptor antagonism. A novel mechanism for the management of obesity. SMI Inaugural Conference on Obesity and Related Disorders, London.

Callahan, P. M., Rowley, H., Viggers, J., Headland, K., Cheetham, S. C., Rowe, W. B., *et al.* (2009). The novel and selective 5-HT_6 receptor antagonist MEM 68753 reduces food intake and body weight in rats maintained on a high fat diet. Keystone Symposium on Obesity: Novel Aspects of the Regulation of Body Weight, Alberta, Abstract No. 242.

Centers for Disease Control and Prevention (CDC). (2009). Differences in prevalence of obesity among Black, White, and Hispanic adults – United States, 2006–2008. *MMWR Morb. Mortal. Wkly. Rep.* **58**, 740–744.

Centers for Disease Control and Prevention (CDC). (2010). U.S. Obesity Trends. Trends by State 1985–2009. Available at http://www.cdc.gov/obesity/data/trends.html.

Connoley, I.P., Liu, Y.L., Frost, I., Reckless, I.P.,Heal, D.J. and Stock, M.J. (1999). Thermogenic effects of sibutramine and its metabolites. *Br. J. Pharmacol.* **126**(6); 1487–1495.

Corwin, R.L. (2004). Binge-type eating induced by limited access in rats does not require energy restriction on the previous day. *Appetite* **42**(2); 139–142.

Corwin, R.L. (2006). Bingeing rats: a model of intermittent excessive behaviour? *Appetite* **46**(1); 11–15.

Daniels, S.R. (2006). The consequences of childhood overweight and obesity. *Future Child* **16**(1); 47–67.

de Castro, J.J., Dias, T., Chambel, P., Carvalheiro, M., Correia, L.G. and Guerreiro, L., *et al.* (2009). A randomized double-blind study comparing the efficacy and safety of orlistat versus placebo in obese patients with mild to moderate hypercholesterolemia. *Rev. Port. Cardiol.* **28**(12); 1361–1374.

Deedwania, P.C. (2004). Metabolic syndrome and vascular disease: is nature or nurture leading the new epidemic of cardiovascular disease? *Circulation* **109**(1); 2–4.

Deutenberg, P., Deurenberg-Yap, M. and Guricci, S. (2002). Asians are different from Caucasians and from each other in their body mass index/body fat percent relationship. *Obes. Rev.* **3**(3); 141–146.

Dickinson, K., North, T.J., Telford, G., Smith, S., Brammer, R. and Jones, R.B., *et al.* (2001). Determination of body composition in conscious adult female Wistar utilising total body electrical conductivity. *Physiol. Behav.* **74**(4–5); 425–433.

Dujovne, C.A., Zavoral, J.H., Rowe, E. and Mendel, C.M. Sibutramine Study Group (2001). Effects of sibutramine on body weight and serum lipids: a double-blind, randomized, placebo-controlled study in 322 overweight and obese patients with dyslipidemia. *Am. Heart J.* **142**(3); 489–497.

Dutton, G.R., Perri, M.G., Stine, C.C., Goble, M. and Van Vessem, N. (2010). Comparison of physician weight loss goals for obese male and female patients. *Prev. Med.* **50**(4); 186–188.

East, S.Z., Burnet, P.W., Leslie, R.A., Roberts, J.C. and Harrison, P.J. (2002). 5-HT_6 receptor binding sites in schizophrenia and following antipsychotic drug administration: autoradiographic studies with [^{125}I]SB-258585. *Synapse* **45**(3); 191–199.

Ello-Martin, J.A., Ledikwe, J.H. and Rolls, B.J. (2005). The influence of food portion size and energy density on energy intake: implications for weight management. *Am. J. Clin. Nutr.* **82**(1); 236S–241S.

European Medicines Agency. (1999). Committee for Medicinal Products for Human Use (CHMP). Press release http://www.ema.europa.eu/docs/en_GB/document_library/Press_release/2009/12/WC500017629.pdf.

European Medicines Agency. (2007). Committee for Medicinal Products for Human Use (CHMP). Guidance on clinical evaluation of medicinal products used in weight control. Available at http://www.ema.europa.eu/docs/en_GB/document_library/Scientific_guideline/2009/09/WC500003264.pdf.

European Medicines Agency. (2010). Press release: European Medicines Agency recommends suspension of marketing authorisations for sibutramine. Weight-loss medicine associated with increased risk of cardiovascular events to be removed from all markets in the European Union. EMA/39408/2010, available at http://www.ema.europa.eu/ema/index.jsp?curl=pages/news_and_events/news/2010/01/news_detail_000985.jsp&mid=WC0b01ac058004d5c1&murl=menus/news_and_events/news_and_events.jsp&jsenabled=true.

Food and Drug Administration (FDA). (2007). Center for Drug Evaluation and Research (CDER). Draft. Guidance for Industry. Developing Products for Weight Management. Available at http://www.fda.gov/downloads/Drugs/GuidanceComplianceRegulatoryInformation/Guidances/ucm071612.pdf.

Food and Drug Administration (FDA). (2009). Early communication about an ongoing safety review of Meridia (sibutramine hydrochloride). Available at http://www.fda.gov/Drugs/DrugSafety/PostmarketDrugSafetyInformationforPatientsandProviders/DrugSafetyInformationforHeathcareProfessionals/ucm191650.htm.

Food and Drug Administration (FDA). (2010). Follow-up to the November 2009 early communication about an ongoing safety review of sibutramine, marketed as Meridia. Available at http://www.fda.gov/Drugs/DrugSafety/PostmarketDrugSafetyInformationforPatientsandProviders/DrugSafetyInformationforHeathcareProfessionals/ucm198206.htm.

Felson, D.T. (1996). Weight and osteoarthritis. *Am. J. Clin. Nutr.* **63** (3 Suppl.); 430S–432S.

Fisas, A., Codony, X., Romero, G., Dordal, A., Giraldo, J. and Merce, R., *et al.* (2006). Chronic 5-HT$_6$ receptor modulation by E-6837 induces hypophagia and sustained weight loss in diet-induced obese rats. *Br. J. Pharmacol.* **148**(7); 973–983.

Flack, J.M., Peters, R., Shafi, T., Alrefai, H., Nasser, S.A. and Crook, E. (2003). Prevention of hypertension and its complications: theoretical basis and guidelines for treatment. *J. Am. Soc. Nephrol.* **14**(Suppl. 2); S92–S98.

Flegal, K.M., Carroll, M.D., Ogden, C.L. and Curtin, L.R. (2010). Prevalence and trends in obesity among US adults, 1999–2008. *J. Am. Med. Assoc.* **303**(3); 235–241.

Florentin, M., Kostapanos, M.S., Nakou, E.S., Elisaf, M. and Liberopoulos, E.N. (2009). Efficacy and safety of ezetimibe plus orlistat or rimonabant in statin-intolerant nondiabetic overweight/obese patients with dyslipidemia. *J. Cardiovasc. Pharmacol. Ther.* **14**(4); 274–282.

Foley, J.E. and Jordan, J. (2010). Weight neutrality with the DPP-4 inhibitor, vildagliptin: mechanistic basis and clinical experience. *Vasc. Health Risk Manag.* **6**, 541–548.

Fone, K. C. F., Marsden, C. A., Bentley, J. C., and Woolley, M. L. (2002). Functional studies on 5-HT$_6$ receptors. Proceedings of the 5th IUPHAR Serotonin Satellite Meeting, Acapulco, Mexico, P77.

Frassetto, A., Zhang, J., Lao, J.Z., White, A., Metzger, J.M. and Fong, T.M., *et al.* (2008). Reduced sensitivity to diet-induced obesity in mice carrying a mutant 5-HT$_6$ receptor. *Brain Res.* **1235**, 140–144.

Fujioka, K., Seaton, T.B., Rowe, E., Jelinek, C.A., Raskin, P. and Lebovitz, H.E., *et al.* (2000). Weight loss with sibutramine improves glycaemic control and other metabolic parameters in obese patients with Type 2 diabetes mellitus. *Diabetes Obes. Metab.* **2**(3); 175–187.

Gannon, K.S., Heal, D.J., Cheetham, S.C., Jackson, H.C., Seeley, H.C. and Melendez, R.J., *et al.* (2006a). PRX-07034, a potent and selective 5-HT_6 receptor antagonist, reduces food intake and body weight in rats. *J. Pharmacol. Sci.* **101**(Suppl. 1); 124.

Gannon, K. S., Shacham, S., Noiman, S., Heal, D., Seeley, R., Melendez, R., *et al.* (2006b). PRX-07034, a potent and selective 5-HT_6 receptor antagonist, for the treatment of obesity. Proceedings of NAASO: The North American Obesity Society Annual Scientific Meeting, Boston.

Gerard, C., el Mestikawy, S., Lebrand, C., Adrien, J., Ruat, M. and Traiffort, E., *et al.* (1996). Quantitative RT-PCR distribution of serotonin 5-HT_6 receptor mRNA in the central nervous system of control or 5,7-dihydroxytryptamine-treated rats. *Synapse* **23**(3); 164–173.

Gerard, C., Martres, M.P., Lefevre, K., Miquel, M.C., Verge, D. and Lanfumey, L., *et al.* (1997). Immuno-localization of serotonin 5-HT_6 receptor-like material in the rat central nervous system. *Brain Res.* **746**(1–2); 207–219.

Gokcel, A., Karakose, H., Ertorer, E.M., Tanaci, N., Tutuncu, N.B. and Guvener, N. (2001). Effects of sibutramine in obese female subjects with Type 2 diabetes and poor blood glucose control. *Diabetes Care* **24**(11); 1957–1960.

Goldstein, B.J., Feinglos, M.N., Lunceford, J.K., Johnson, J. and Williams-Herman, D.E. Sitagliptin 036 Study Group (2007). Effect of initial combination therapy with sitagliptin, a dipeptidyl peptidase-4 inhibitor, and metformin on glycemic control in patients with Type 2 diabetes. *Diabetes Care* **30**(8); 1979–1987.

Goodman, E. and Whitaker, R.C. (2002). A prospective study of the role of depression in the development and persistence of adolescent obesity. *Pediatrics* **110**(3); 497–504.

Greenway, F. L., Fujioka, K., Gupta, A. K., Smith, S. R., Guttadauria, M., Tollefson, G., *et al.* (2007). A double-blind, placebo-, bupropion- and naltrexone-controlled study of the efficacy and safety of three doses of naltrexone-bupropion SR combination therapy in obesity: effects on total and visceral adipose tissue and CV risk markers. Proceedings of NAASO: The North American Obesity Society Annual Scientific Meeting, New Orleans. Poster available at http://www.vivus.com.

Haddock, C.K., Poston, W.S.C., Dill, P.L., Foreyt, J.P. and Ericsson, M. (2002). Pharmacotherapy for obesity: a quantitative analysis of four decades of published randomised clinical trials. *Int. J. Obes. Relat. Metab. Disord.* **26**(2); 262–273.

Halford, J.C., Boyland, E.J., Cooper, S.J., Dovey, T.M., Huda, M.S. and Dourish, C.T., *et al.* (2010). The effects of sibutramine on the microstructure of eating behaviour and energy expenditure in obese women. *J. Psychopharmacol.* **24**(1); 99–109.

Hamon, M., Doucet, E., Lefèvre, K., Miquel, M.-C., Lanfumey, L. and Insausti, R., *et al.* (1999). Antibodies and antisense oligonucleotide for probing the distribution and putative functions of central 5-HT_6 receptors. *Neuropharmacology* **21**(Suppl. 2); 68S–76S.

Heal, D.J., Aspley, S., Prow, M.R., Jackson, H.C., Martin, K.F. and Cheetham, S.C. (1998). Sibutramine: a novel anti-obesity drug. A review of the pharmacological evidence to differentiate it from d-amphetamine and d-fenfluramine. *Int. J. Obes. Relat. Metab. Disord.*S18–S28 discussion S29

Heal, D.J., Gosden, J. and Smith, S.L. (2009). Regulatory challenges for new drugs to treat obesity and comorbid metabolic disorders. *Br. J. Clin. Pharmacol.* **68**(6); 861–874.

Heal, D.J. and Jagger, E. (2005). Development, regulatory and marketing challenges for novel anti-obesity therapies. In: Antel, J., Finer, N., Heal, D., Krause, G. (Eds.), *Obesity and Metabolic Disorders*. IOS Press, Amsterdam, pp. 73–91.

Heal, D.J., Smith, S.L., Fisas, A., Codony, X. and Buschmann, J. (2008). Selective 5-HT_6 receptor ligands: progress in the development of a novel pharmacological approach to the treatment of obesity and related metabolic disorders. *Pharmacol. Ther.* **117**(2); 207–231.

Heshka, S. and Heymsfield, S. (2001). Obesity and gallstones. In: Bjorntorp, P. (Ed.), *International Textbook of Obesity*. Wiley, Chichester, pp. 399–409.

Heymsfield, S.B., Segal, K.R., Hauptman, J., Lucas, C.P., Boldrin, M.N. and Rissanen, A., *et al.* (2000). Effects of weight loss with orlistat on glucose tolerance and progression to Type 2 diabetes in obese adults. *Arch. Intern. Med.* **160**(9); 1321–1326.

Hirst, W.D., Abrahamsen, B., Blaney, F.E., Calver, A.R., Aloj, L. and Price, G.W., *et al.* (2003). Differences in the central nervous system in distribution and pharmacology of the mouse 5-hydroxytryptamine-6 receptor compared with rat and human receptors investigated by radioligand binding, site-directed mutagenesis, and molecular modelling. *Mol. Pharmacol.* **64**(6); 1295–1308.

Hollander, P.A., Elbein, S.C., Hirsch, I.B., Kelley, D., McGill, J. and Taylor, T., *et al.* (1998). Role of orlistat in the treatment of obese patients with Type 2 diabetes. A 1-year randomized double-blind study. *Diabetes Care* **21**(8); 1288–1294.

International Obesity Task Force [IOTF]. (2010). Available at http://www.iotf.org.

Jackson, H. C., Cheetham, S. C., Dickinson, K., Jones, R. B., Heal, D. J., Gregory, P., *et al.* (2005). Comparison of the effects of rimonabant and sibutramine in a rat model of dietary-induced obesity. Society for Neurosciences Meeting, Abstract 532.13. Available at www.sfn.org.

Jackson, H. C., Cheetham, S. C., Gregory, P. C., and Antel, J. (2007). Effect of chronic administration of topiramate and phentermine alone and in combination in an animal model of dietary induced obesity. Society for Neurosciences Meeting, Abstract 629.15. Available at www.sfn.org.

Jackson, H. C., Dickinson, K., Jones, R. B., Schumacher, C., and Jensen, C. (2004). Comparison of the effects of sibutramine and orlistat on body weight, food and water intake and body composition in dietary-induced obese rats. Society for Neurosciences Meeting, Abstract 75.3. Available at www.sfn.org.

Jackson, H.C., Needham, A.M., Hutchins, L.J., Mazurkiewicz, S.E. and Heal, D.J. (1997). Comparison of the effects of sibutramine and other monoamine reuptake inhibitors on food intake in the rat. *Br. J. Pharmacol.* **121**(8); 1758–1762.

James, W.P.T., Astrup, A., Finer, N., Hilsted, J., Kopelman, P. and Rössner, S., *et al.* (2000). Effect of sibutramine on weight maintenance after weight loss: a randomised trial. *Lancet* **356**(9248); 2119–2125.

James, W.P.T., Chunming, C. and Inoue, S. (2002). Appropriate Asian body mass indices? *Obes. Rev.* **3**(3); 137–232.

Jones, R. B., Jackson, H. C., Dickinson, D., Schumacher, C., and Jensen, C. (2005). Comparison of the effects of sibutramine and orlistat on body weight, food and water intake and body composition in dietary-induced obese rats. Poster at IBC's 3rd Annual Targeting Metabolic Syndrome Conference, Cambridge, USA.

Kelley, D.E., Bray, G.A., Pi-Sunyer, F.X., Klein, S., Hill, J. and Miles, J., *et al.* (2002). Clinical efficacy of orlistat therapy in overweight and obese patients with insulin-treated Type 2 diabetes: a 1-year randomized controlled trial. *Diabetes Care* **25**(6); 1033–1041.

Kim, D. (2010). Contrave: weight loss, cardiometabolic improvements, and better control of eating. Oral communication: CBI's 6th Annual Obesity and Diabetes Drug Development Summit, Arlington, VA.

Kohen, R., Metcalf, M.A., Khan, N., Druck, T., Huebner, K. and Lachowicz, J.E., *et al.* (1996). Cloning, characterization, and chromosomal localization of a human 5-HT_6 serotonin receptor. *J. Neurochem.* **66**(1); 47–56.

Law, M., Wald, N. and Morris, J. (2003). Lowering blood pressure to prevent myocardial infarction and stroke: a new preventive strategy. *Health Technol. Assess.* **7**(31); 1–94.

Li, Z., Maglione, M., Tu, W., Mojica, W., Arterburn, D. and Shugarman, L.R., *et al.* (2005). Meta-analysis: pharmacologic treatment of obesity. *Ann. Intern. Med.* **142**(7); 532–546.

Linné, Y., Hemmingsson, E., Adolfsson, B., Ramsten, J. and Rössner, S. (2002). Patient expectations of obesity treatment - the experience from a day-care unit. *Int. J. Obes. Relat. Metab. Disord.* **26**(5); 739–741.

Marcos, B., Gil-Bea, F.J., Hirst, W.D., Garcia-Alloza, M. and Ramirez, M.J. (2006). Lack of localization of 5-HT_6 receptors on cholinergic neurons: implication of multiple neurotransmitter systems in 5-HT_6 receptor-mediated acetylcholine release. *Eur. J. Neurosci.* **24**(5); 1299–1306.

Max, S.I., Monsma Jr., F.J. and Sibley, D.R. (1995). Agonist-induced desensitization of 5-HT_6 serotonin receptor-coupled adenylyl cyclase in stably transfected HEK-293 cells. *J. Serotonin Res.* **2**, 101–116.

McNeely, W. and Benfield, P. (1998). Orlistat. *Drugs* **56**(2); 241–249.

McNulty, S.J., Ur, E. and Williams, G. Multicenter Sibutramine Study Group (2003). A randomized trial of sibutramine in the management of obese Type 2 diabetic patients treated with metformin. *Diabetes Care* **26**(1); 125–131.

Metcalf, B. S., Hosking, J., Jeffery, A. N., Voss, L. D., Henley, W., and Wilkin, T. J. (2010). Fatness leads to inactivity, but inactivity does not lead to fatness: a longitudinal study in children (Earlybird 45). *Arch. Dis. Child.* **June 23** (Epub ahead of print).

Mokdad, A.H., Bowman, B.A., Ford, E.S., Vinicor, F., Marks, J.S. and Koplan, J.P. (2001). The continuing epidemics of obesity and diabetes in the United States. *J. Am. Med. Assoc.* **286**(10); 1195–1200.

Monsma Jr., F.J., Shen, Y., Ward, R.P., Hamblin, M.W. and Sibley, D.R. (1993). Cloning and expression of a novel serotonin receptor with high affinity for tricyclic psychotropic drugs. *Mol. Pharmacol.* **43**(3); 320–327.

Montani, J.-P., Antic, V., Yang, Z. and Dulloo, A. (2002). Pathways from obesity to hypertension: from the perspective of a vicious triangle. *Int. J. Obes.* **26**(Suppl. 2); S28–S38.

Moon, O.R., Kim, N.S., Jang, S.M., Yoon, T.H. and Kim, O. (2002). The relationship between body mass index and the prevalence of obesity-related diseases based on the 1995 National Health Interview Survey in Korea. *Obes. Rev.* **3**(3); 191–196.

Murray, S. R., Callahan, P. M., Rowe, W. B., Tehim, A., Murray, M., Wang, S., *et al.* (2008). MEM 68626, a selective serotonin 5-HT_6 receptor antagonist as a potential treatment for Alzheimer's disease, mild cognitive impairment and obesity. Proceedings of the American College of Neuropsychopharmacology (ACNP) 47th Annual Meeting, Scottsdale.

National Cholesterol Education Program. (2002). Third report of the National Cholesterol Education Program (NCEP) expert panel on detection, evaluation, and treatment of high blood cholesterol in adults (Adult Treatment Panel III) Final report. *Circulation,* **106**, 3143–3420.

Nielsen, S. and Jensen, M.D. (1997). Obesity and cardiovascular disease: is body structure a factor? *Curr. Opin. Lipidol.* **8**(4); 200–204.

Nissen, S.E., Nicholls, S.J., Wolski, K., Nesto, R., Kupfer, S. and Perez, A., *et al.* (2008). Comparison of pioglitazone vs. glimepiride on progression of coronary atherosclerosis in patients with Type 2 diabetes: the PERISCOPE randomized controlled trial. *J. Am. Med. Assoc.* **299**(13); 1561–1573.

Peiris, A.N., Sothmann, M.S., Hoffmann, R.G., Hennes, M.I., Wilson, C.R. and Gustafson, A.B., *et al.* (1989). Adiposity, fat distribution, and cardiovascular risk. *Ann. Intern. Med.* **110**(11); 867–872.

Pi-Sunyer, X. (2002). The obesity epidemic: pathophysiology and consequences of obesity. *Obes. Res.* **10**(Suppl. 2); 97S–104S.

Prentice, A.M. and Jebb, S.A. (1995). Obesity in Britain: gluttony or sloth? *Br. Med. J.* **311**(7002); 437–439.

Rashid, P., Leonardi-Bee, J. and Bath, P. (2003). Blood pressure reduction and secondary prevention of stroke and other vascular events: a systematic review. *Stroke* **34**(11); 2741–2749.

Richelsen, B., Tonstad, S., Rössner, S., Toubro, S., Niskanen, L. and Madsbad, S., *et al.* (2007). Effect of orlistat on weight regain and cardiovascular risk factors following a very-low-energy diet in abdominally obese patients: a 3-year randomized, placebo-controlled study. *Diabetes Care* **30**(1); 27–32.

Roberts, J.C., Reavill, C., East, S.Z., Harrison, P.J., Patel, S. and Routledge, C., *et al.* (2002). The distribution of 5-HT_6 receptors in rat brain: an autoradiographic binding study using the radiolabelled 5-HT_6 receptor antagonist [^{125}I]SB-258585. *Brain Res.* **934**(1); 49–57.

Ruat, M., Traiffort, E., Arrang, J.M., Tardivel-Lacombe, J., Diaz, J. and Leurs, R., *et al.* (1993). A novel rat serotonin (5-HT$_6$) receptor: molecular cloning, localization and stimulation of cAMP accumulation. *Biochem. Biophys. Res. Commun.* **193**(1); 268–276.

Saraç, F., Pehlivan, M., Çelebi, G., Saygili, F., Yilmaz, C. and Kabalak, T. (2006). Effects of sibutramine on thermogenesis in obese patients assessed via immersion calorimetry. *Adv. Ther.* **23**(6); 1016–1029.

Sargent, B. and Moore, N.A. (2009). New central targets for the treatment of obesity. *Br. J. Clin. Pharmacol.* **68**(6); 852–860.

Sastry, K. R., Vishwakarma, S., Jayarajan, P., Abraham, R., Koul, S., Khan, F., *et al.* (2007). Centrally acting anti-obesity agent SUVN-504 – a potent 5-HT$_6$ receptor antagonist. Society for Neurosciences Meeting, Abstract 300.23. Available at www.sfn.org.

Setola, V. and Roth, B.L. (2003). Why mice are neither miniature humans nor small rats: a cautionary tale involving 5-hydroxytryptamine-6 serotonin receptor species variants. *Mol. Pharmacol.* **64**(6); 1277–1278.

Shacham, S., Heal, D.J., Cheetham, S. C., Jackson, H. C., Melendez, R., Sharadendu, A., *et al.* (2006). PRX-07034, a potent and selective 5-HT$_6$ receptor antagonist, reduces food intake and body weight in dietary-induced obese rats. Society for Neurosciences Meeting, Abstract 62.10. Available at www.sfn.org.

Shah, P. and Mudaliar, S. (2010). Pioglitazone: side effect and safety profile. *Expert Opin. Drug Saf.* **9**(2); 347–354.

Shanmuganathan, D., Koul, S., Abraham, R., Khan, F., Jayarajan, P., and Nirogi, R. (2008). SUVN-51005 – a potent 5-HT$_6$ antagonist in animal models of obesity. Society for Neurosciences Meeting, Abstract 584.18. Available at www.sfn.org.

Simon, G.E., Von Korff, M., Saunders, K., Miglioretti, D.L., Crane, P.K. and van Belle, G., *et al.* (2006). Association between obesity and psychiatric disorders in the US adult population. *Arch. Gen. Psychiatry* **63**(7); 824–830.

Smith, S.R., Prosser, W.A., Donahue, D.J., Morgan, M.E., Anderson, C.M. and Shanahan, W.R. APD356-004 Study Group (2009). Lorcaserin (APD356), a selective 5-HT$_{2C}$ agonist, reduces body weight in obese men and women. *Obesity* **17**(3); 494–503.

Smith, S.R., Weissman, N.J., Anderson, C.M., Sanchez, M., Chuang, E. and Stubbe, S., *et al.* Behavioral Modification and Lorcaserin for Overweight and Obesity Management (BLOOM) Study Group (2010). Multicenter, placebo-controlled trial of lorcaserin for weight management. *N. Engl. J. Med.* **363**(3); 245–256.

Svartengren, J., Axelsson-Lendin, P., Edling, N., Fhölenhag, K., Isacson, R., Hillegaart, V., *et al.* (2004). The selective serotonin 5-HT$_6$ receptor antagonist BVT.5182 decreases food intake and body weight in both rats and mice. Society for Neurosciences Meeting, Abstract 75.8. Available at www.sfn.org.

Svartengren, J., Axelsson-Lendin, P., Multan, U., Berggren, V., Isacson, R., Hillegaart, V., *et al.* (2007). Characterization of a new selective BVT serotonin 5-HT$_6$ receptor antagonist in animal models for feeding and body weight regulation. Society for Neurosciences Meeting, Abstract 629.11. Available at www.sfn.org.

Svartengren, J., Ohman, B., Edling, N., Svensson, M., Fhölenhag, M., Axelsson-Lendin, P., *et al.* (2003). The serotonin 5-HT$_6$ receptor antagonist BVT.5182 reduces body weight of high fat diet-induced mice. Proceedings of the European Congress on Obesity (ECO) Meeting, Finland. Poster P1-094.

Tecott, L. H., and Brennan, T. J. (2000). Serotonin 5-HT$_6$ receptor knockout mouse. US Patent 6060642. Available at http://www.freepatentsonline.com/6060642.html.

Torgerson, J.S., Hauptman, J., Boldrin, M.N. and Sjöström, L. (2004). XENical in the prevention of diabetes in Obese Subjects (XENDOS) study: a randomized study of orlistat as an adjunct to lifestyle changes for the prevention of Type 2 diabetes in obese patients. *Diabetes Care* **27**(1); 155–161.

UK Prospective Diabetes Study Group (1998). Tight blood pressure control and risk of macrovascular and microvascular complications in Type 2 diabetes: UKPDS 38. *Br. Med. J.* **317**, 703–713.

UK Prospective Diabetes Study Group (2004). Risk factors for myocardial infarction case fatality and stroke case fatality in Type 2 diabetes. UKPDS 66. *Diabetes Care* **27**, 201–207.

Upton, N., Chuang, T.T., Hunter, A.J. and Virley, D.J. (2008). 5-HT$_6$ receptor antagonists as novel cognitive enhancing agents for Alzheimer's disease. *Neurotherapeutics* **5**(3); 458–469.

Vivus. (2009). Qnexa Phase 3 Results Conference Call. Equip and Conquer Study Results. Available at http://www.vivus.com/images/stories/downloads_newsroom/qnexa982009.pdf.

Wadden, T.A., Berkowitz, R.I., Womble, L.G., Sarwer, D.B., Phelan, S. and Cato, R.K., *et al.* (2005). Randomized trial of lifestyle modification and pharmacotherapy for obesity. *N. Engl. J. Med.* **353** (20); 2111–2120.

Ward, R.P., Hamblin, M.W., Lachowicz, J.E., Hoffman, B.J., Sibley, D.R. and Dorsa, D.M. (1995). Localization of serotonin subtype 6 receptor messenger RNA in the rat brain by in situ hybridization histochemistry. *Neuroscience* **64**(4); 1105–1111.

Wardle, J., Waller, J. and Fox, E. (2002). Age of onset and body dissatisfaction on obesity. *Addict. Behav.* **27**(4); 561–573.

Wojnicki, F.H.E., Johnson, D.S. and Corwin, R.L.W. (2008). Access conditions affect binge-type shortening consumption in rats. *Physiol. Behav.* **95**(5); 649–657.

Wolk, R., Shamsuzzaman, A.S. and Somers, V.K. (2003). Obesity, sleep apnoea and hypertension. *Hypertension* **42**(6); 1067–1074.

Woolley, M.L., Bentley, J.C., Sleight, A.J., Marsden, C.A. and Fone, K.C.F. (2001). A role for 5-HT$_6$ receptors in retention of spatial learning in the Morris water maze. *Neuropharmacology* **41**(2); 210–219.

Woolley, M.L., Marsden, C.A. and Fone, K.C. (2004). 5-HT$_6$ receptors. *Curr. Drug Targets CNS Neurol. Disord.* **3**(1); 59–79.

BEHAVIORAL AND NEUROCHEMICAL PHARMACOLOGY OF 5-HT_6 RECEPTORS RELATED TO REWARD AND REINFORCEMENT

Gaetano Di Chiara[1,2,3,4], Valentina Valentini[1,2,3], and Sandro Fenu[1,2,3]

[1]Department of Toxicology, [2]National Institute of Neuroscience, [3]Centre of Excellence "Neurobiology of Addiction", University of Cagliari, Italy
[4]CNR Institute of Neuroscience, section of Cagliari, Italy

I. Introduction

Much of the initial interest on 5-HT_6 receptors has arisen from their specific distribution to the brain, their high density in the striatum, including the resemblance with the distribution of dopamine (DA) receptors, and the high affinity displayed by atypical antipsychotics. The distribution of 5-HT_6 receptors in the brain predicted a role in reward, motivation and extrapyramidal motor functions, and an interaction with DA. Indeed, this prediction has not been met until recently. Still, the evidence for a role of 5-HT_6 receptors in motivation, reward,

INTERNATIONAL REVIEW OF
NEUROBIOLOGY, VOL. 96
DOI: 10.1016/B978-0-12-385902-0.00005-X

and extrapyramidal functions is scanty and contradictory. However, it is possible that some information on the neurochemical and behavioral effects of 5-HT$_6$ receptor agonists and antagonists might be relevant for an involvement of these receptors in those functions. Because of this, we will provide a broad review of the neurochemical and behavioral pharmacology of 5-HT$_6$ receptors rather than one strictly confined to their role in motivation, reward, and extrapyramidal functions.

II. Distribution of 5-HT$_6$ Receptors

A. Striatal Areas

The density of 5-HT$_6$ receptors is particularly high in the striatal areas (caudate-putamen (CPu), nucleus accumbens (NAc), and olfactory tubercle) of rats and man. In the rat, the olfactory tubercle is the area with the highest signal for 5-HT$_6$ receptor mRNA as estimated by *in situ* hybridization (Ballaz *et al.*, 2007; Ward *et al.*, 1995). Next is, in the rat, the NAc and the CPu with a dorsoventral gradient of increasing density (Ward *et al.*, 1995). Using RT-PCR and ligand binding in homogenates, allowing better quantitation but lower topographic resolution, the density of 5-HT$_6$ receptors and the expression of mRNA in the CPu of rat and man are about 10 times higher than those in nonstriatal areas (Hirst *et al.*, 2003). Simple visual inspection of autoradiographic images of the specific binding of the 5-HT$_6$ receptor antagonist ^{125}I-SB-258585 in coronal brain slices, taken at the level of the rat basal ganglia, shows a striking similarity with the distribution of specific DA receptor binding (Roberts *et al.*, 2002).

Electron microscopic studies of immunolabeled striatal 5-HT$_6$ receptors show a postsynaptic localization on dendrites and labeling of postsynaptic specializations (Gerard *et al.*, 1997; Hamon *et al.*, 1999). Light microscopic examination of immunolabeled slices at high magnification shows 5-HT$_6$-like immunoreactivity on microtubular structures called “cilia” (Brailov *et al.*, 2000; Hamon *et al.*, 1999).

5-HT$_6$ receptor mRNA is expressed by putative medium-size spiny neurons with no selectivity among neurons expressing substance P/dynorphin and projecting to the substantia nigra/internal pallidum and neurons expressing enkephalin and projecting to the external pallidum (Ward and Dorsa, 1996). 6-OHDA lesions of DA neurons do not affect the binding of 5-HT$_6$ ligand estimated by autoradiography (Roberts *et al.*, 2002), indicating that 5-HT$_6$ receptors are not located on DA neurons.

In spite of the evidence of the role of acetylcholine (ACh) in some behavioral effects of 5-HT$_6$ receptor antagonists (see below), no double-labeling studies investigating the relationship between 5-HT_6 receptors and striatal ACh neurons are available.

B. Extrastriatal Areas

A high level of expression of mRNA for the 5-HT_6 receptor, similar to that observed in the striatum, is found in the granular layer of the dentate girus and CA fields of the hippocampus of the rat (Ballaz *et al.*, 2007; Ward *et al.*, 1995). Moreover, detailed analysis of 5-HT_6 receptor distribution by autoradiographic binding studies in the rat shows moderately high levels in the hippocampus and neocortex; central, paraventricular, and reticular thalamic nuclei; selected hypothalamic nuclei; and habenula (Ballaz *et al.*, 2007; Roberts *et al.*, 2002). Binding of the 5-HT_6 ligand was high also in the choroid plexus (Roberts *et al.*, 2002). In contrast to studies utilizing antibodies against the receptor protein (Gerard *et al.*, 1997; Hamon *et al.*, 1999), binding (Hirst *et al.*, 2003; Roberts *et al.*, 2002) as well as mRNA expression studies (Gerard *et al.*, 1997; Hirst *et al.*, 2003; Ruat *et al.*, 1993; Ward and Dorsa, 1996; Ward *et al.*, 1995) show very low levels of 5-HT_6 receptors in the cerebellum. Therefore, detailed analysis of 5-HT_6 receptor distribution indicates a widespread function of these receptors beyond that of the striatum and of DA.

The 5-HT_6 receptor mRNA signal has not been detected in nonnervous peripheral tissues; therefore, this receptor is the only 5-HT receptor to be expressed selectively in the nervous system (Hirst *et al.*, 2003; Hoyer *et al.*, 1994; Monsma *et al.*, 1993; Ruat *et al.*, 1993).

C. Species Specificity

An important aspect of the 5-HT_6 receptor neurobiology is its species specificity. Thus, in contrast to rats and men, the expression of 5-HT_6 receptors in the mouse brain is very low (Hirst *et al.*, 2000, 2003); moreover, the mouse receptor itself has different pharmacological properties from the rat and the human receptors. Thus, the selective antagonist Ro 04-6790 does not bind to the mouse receptor and another antagonist SB-258585 has an affinity about two orders of magnitude lower for the mouse than for the rat and human receptors, which, instead, have similar pharmacological properties, density, and distribution (Hirst *et al.*, 2003). On the basis of site-directed mutagenesis studies, these differences have been attributed to a difference in two amino acids placed in a critical position: Tyr-188 in place of Phe 188 and Ser 290 in place of Asn 190 (Hirst *et al.*, 2003).

These differences cast some doubt on the relevance of experimental studies performed in mice for the role of 5-HT_6 receptors in humans.

III. Signal Transduction

5-HT$_6$ receptor transduction was originally shown to take place by stimulation of adenylate cyclase via a GαS protein in transfected cells expressing the cloned receptor (Hirst *et al.*, 2003; Monsma *et al.*, 1993; Ruat *et al.*, 1993). Other mammalian 5-HT receptors coupled to adenylate cyclase are the 5-HT$_7$ and 5-HT$_4$ receptors. While 5-HT$_4$ receptors are fairly abundant in striatal areas, 5-HT$_7$ receptors are poorly expressed there.

The native 5-HT$_6$ receptor can also bind to two other proteins that might be part of alternative transduction pathways of the 5-HT$_6$ signal. These proteins are Fyn and Jab1 (Yun *et al.*, 2007, 2010). Fyn is a protein with tyrosine kinase activity, regulated by phosphorylation of Tyr-420, which results in kinase activation, and Tyr-531, resulting in inhibition. The 5-HT$_6$ receptor, via its carboxyl terminal, binds to the SH3 region of Fyn, thus inducing Tyr-420 autophosphorylation and full activation of kinase activity. This activity in turn, via a cascade of tyrosine kinases (Ras and MEK-dependent pathways), finally activates ERK 1/2. Fyn and 5-HT$_6$ receptor interactions reciprocally facilitate their functions. Thus, 5-HT$_6$ receptor activation promotes translocation of Fyn from the nucleus to the cytoplasm, by binding it to the internal side of the plasma membrane through its carboxyl terminal. On the other hand, increased Fyn expression increases 5-HT$_6$ receptor expression (Yun *et al.*, 2007). Jab1 is a Jun activation domain binding protein that also can bind to the 5-HT$_6$ receptor and whose silencing results in a decrease of 5-HT$_6$ receptor expression. Activation of 5-HT$_6$ receptors promotes translocation of Jab1 to the nucleus and increased c-Jun phosphorylation (Yun *et al.*, 2010).

These different pathways, particularly the adenylate cyclase, via PKA phosphorylation of CREB and the Fyn pathway, can converge into ERK phosphorylation.

IV. Interaction of Antipsychotics with 5-HT$_6$ Receptors

The discovery of 5-HT$_6$ receptors attracted much interest not only because their high striatal density distribution but also because clozapine, an atypical antipsychotic, showed nanomolar affinity for the recombinant rat 5-HT$_6$ receptor (Monsma *et al.*, 1993). Roth *et al.* (1994) showed that a number of typical neuroleptics, including chlorprotixene, chlorpromazine, amoxapine, thioridazine (a neuroleptic with an atypically high antimuscarinic activity), loxapine, clothiapine, fluphenazine, and perphenazine, and atypical neuroleptics, including zotepine, olanzapine, clozapine, fluperlapine, and tiospirone, had K_i lower than 20 nM. Glatt *et al.* (1995) showed that the high-affinity binding profile of ^{3}H-clozapine to rat brain membranes (K_d 4.5 nM) was that of a 5-HT_6 receptor ligand. Boess *et al.* (1997) and Hirst *et al.* (2003) confirmed the high-affinity binding of clozapine to rat and human 5-HT_6 receptors using ^{3}H-LSD and ^{3}H-5-HT as ligands and Boess *et al.* (1998) confirmed the binding to rat and pig 5-HT_6 receptors with ^{3}H-Ro 63-0563. Recently, Dupuis *et al.* (2008) have tested 24 antipsychotics on a scintillation proximity assay of coupling to GαS protein and showed that not only clozapine but also its major metabolite, *N*-desmethylclozapine, bound with nanomolar affinity to 5-HT_6 receptors and behaved as antagonists. Haloperidol, quetiapine, and risperidone had micromolar affinity (Dupuis *et al.*, 2008). These observations indicate that the ability to block 5-HT_6 receptors is not specifically associated with the atypical properties of the antipsychotics and tends to exclude that these receptors play a role in the extrapyramidal side effects of antipsychotics.

Consistent with an interaction of clozapine with 5-HT_6 receptors is the observation by Frederick and Meador-Woodruff (1999) that repeated administration of clozapine (20 mg/kg) in the rat decreased the expression of mRNA for 5-HT_6 receptors specifically in all four CA subfields and in the dentate girus of the hippocampus, while administration of haloperidol (2 mg/kg) was ineffective.

Although the ability to block 5-HT_6 receptors does not seem to be involved in the extrapyramidal side effects of antipsychotics, the 5-HT_6 receptor blocking property of some antipsychotics might add to their profile some peculiar cognitive or emotional properties.

In relation to this, Li *et al.* (2007) have reported that the 5-HT_6 antagonist SB-399885 potentiated the ability of haloperidol and of the atypical antipsychotic risperidone in the rat, both devoid of 5-HT_6 receptor blocking properties, to increase DA in dialysates of the hippocampus; SB-399885 potentiated the DA-releasing properties of risperidone also in the medial prefrontal cortex. These observations suggest that the property of blocking 5-HT_6 receptors contributes to the DA-releasing properties of antipsychotics in allocortical and periallocortical areas. This effect, in turn, has been suggested to exert a positive action on the cognitive symptoms of schizophrenia (Tan *et al.*, 2009).

V. Behavior- and Drug-Related Changes in 5-HT_6 Receptor Expression

Sensation seeking is a normal, genetically influenced personality trait that can bring to a disorder of impulsivity characterized by risky behavior, including drug seeking and dependence. This trait has been modeled in animals by locomotion in response to a novel environment (novelty seeking) (Bardo *et al.*, 1996; Cain *et al.*, 2005; Dellu *et al.*, 1996). In an outbred rat population, the extremes in the distribution curve for individual locomotor response to novelty (high responder

[HR] and low responder [LR]) are characterized by differential responses to the locomotor stimulant and reinforcing properties of drugs of abuse (Hooks *et al.*, 1991; Piazza *et al.*, 1989, 2000).

Recently, an HR and a LR line has been obtained by selective breeding for responding to novelty (Stead *et al.*, 2006). In these lines, an analysis of mRNA expression by *in situ* hybridization of 5-HT_6 and 5-HT_7 expression shows a 71% higher expression in the HR line compared with the LR in the posterior dentate gyrus of the hippocampus and a 50% decrease in the anterior portion of the same area. In the striatal complex, a significant change (27% decrease) was observed only in the olfactory tubercle and was significantly correlated to the locomotor response to novelty (negative correlation) (Ballaz *et al.*, 2007). 5-HT_7 receptors decreased in the HR compared to LR in the hippocampus and intralaminar and paraventricular thalamus and this was negatively correlated with individual reactivity to novelty (Ballaz *et al.*, 2007). No significant difference were observed in 5-HT_3 receptor expression.

The lack of line-related differences in 5-HT_6 receptor expression in striatal and prefrontal cortical areas traditionally related to impulsivity suggests that the differences observed are related to other differential behavioral and cognitive properties of the two lines.

Kindlundh-Hogberg *et al.* (2006) have studied by RT-PCR the expression of mRNA for various 5-HT receptors including 5-HT_6 receptors in rats administered with MDMA (3×1.0 and 3×5.0 mg/kg/day) every week for 4 weeks. 5-HT_6 mRNA was increased in the frontal cortex and the amygdala. No changes were detected in the CPu, NAc, hippocampus, and hypothalamus (Kindlundh-Hogberg *et al.*, 2006). In the striatum, only mRNA for 5-HT_{1B} receptors was increased, while mRNA for 5-HT_{1B} and 5-HT_{2C} and for 5-HT_3 receptors was increased in the hypothalamus. The relation between these changes and behavior is unknown as no behavioral end point was correlated with the changes in mRNA.

VI. Neurochemical Effects of 5-HT Agonists and Antagonists

The topic of the neurochemical effect of 5-HT_6 receptor agonists and antagonists is controversial. We shall analyze separately the effect of 5-HT_6 antagonists and agonists.

A. 5-HT_6 Antagonists

In the first microdialysis study on this topic, Dawson *et al.* (2000) studied the effect of the 5-HT_6 receptor antagonist SB-271046 on extracellular concentrations of DA, 5-HT, noradrenaline (NE), glutamate, and aspartate in the rat CPu and frontal cortex. SB-271046, given s.c. at doses of 10 mg/kg, increased glutamate and aspartate in the cortex. This effect was delayed, steadily increasing up to maximal levels between 250 and 300 min after drug administration. Perfusion with Ringer containing tetrodotoxin (TTX) brought down the increase in glutamate and aspartate, indicating its dependence upon fast Na channels. Changes in monoamines and amino acids were observed neither in the striatum nor in the cortex.

These observations were confirmed and extended by the same group to the hippocampus, where a TTX-dependent, atropine-independent increase in glutamate, but not in DA and NE, was observed, and to the NAc, where, as in the striatum, no change in monoamines nor glutamate was observed (Dawson *et al.*, 2001). Although no explicit distinction was made between core and shell of the Nac, the reported stereotaxic coordinates suggest that probes were aimed at the NAc core (Dawson *et al.*, 2001).

The dialysate data reported in the above-mentioned studies are characterized by a high degree of variability as indicated by high standard errors and highly scattered changes in dialysate levels of transmitter over time. In view of this, the observations of Dawson *et al.* require independent confirmation.

Lacroix *et al.* (2004) reported an increase in dialysate DA and NA in the medial prefrontal cortex of the rat after 10 mg/kg p.o. of the antagonist SB-271046. Perfusion was made with a Ringer containing neostigmine that, according to the authors, is devoid of any influence on monoami release.

In principle, these observations are not in contrast with those of Dawson *et al.* (2000, 2001), since the area dialyzed in these studies is the lateral frontal cortex, a DA-poor neocortical area, rather than the infralimbic, medial prefrontal cortex, a DA-rich periallocortical area.

However, Li *et al.* (2007), failed to observe any change in medial prefrontal cortex DA dialysates in rats administered with the antagonist SB-399885 given up to doses of 10 mg/kg s.c. At these doses, the antagonist is effective on a variety of behavioral tests (Hirst *et al.*, 2006). Recently, we studied the effect of two 5-HT$_6$ antagonists, SB-271046 (10–20 mg/kg i.p. and 10 mg/kg i.v.) and SB-399885 (5 mg/kg i.p.), on dialysate DA, NA, and 5-HT in the medial prefrontal cortex, NAc shell, and core of the rat and failed to observe any significant change for up to 2 h postdrug (Valentini *et al.*, 2010), in agreement with the observations of Li *et al.* (2007).

As far as ACh is concerned, Shirazi-Southall *et al.* (2002) reported that Ro 04-6790 given i.p. moderately (+50% of basal) increased ACh in dialysates from the rat hippocampus. Moreover, in the studies on the rat, Riemer *et al.* (2003) reported that the 5-HT$_6$ antagonist 4-(2-bromo-6-pyrrolidin-1-ylpyridine-4-sulfonyl)phenylamine (Ro 66-0074) increases ACh in dialysates from the rat frontal cortex and Hirst *et al.* (2006) reported that SB-399885 (10 mg/kg p.o.) increased dialysate ACh in the rat medial prefrontal cortex. *In vitro* studies by Marcos *et al.* (2006) also show that the 5-HT$_6$ antagonist SB-357134 potentiates K^+-induced ^{3}H-ACh release from hippocampal (ED_{50}: 50 nM) and from striatal slices (ED_{50}: 10 nM). These observations suggest an interaction between 5-HT_6 receptors and ACh transmission, although the high sensitivity of prefrontal and hippocampal ACh transmission to arousal state suggests the possibility that these effects are indirect in nature. Thus, an increase in prefrontal ACh transmission might be related to a stimulant effect of the antagonist on arousal. This possibility is consistent with the absence of 5-HT_6 receptors on ACh neurons projecting to the cortex from basal forebrain nuclei, as indicated by the failure of IG-saporin destruction of these neurons to reduce the specific binding of 5-HT_6 ligand in the cortex (Marcos *et al.*, 2006).

However, no information is available on the effect of agonists and antagonists of 5-HT_6 receptors on *in vivo* ACh transmission in the striatum, where an interaction between 5-HT_6 antagonists and muscarinic antagonists has been reported (Bourson *et al.*, 1998).

B. 5-HT_6 Agonists

Only one published study deals with the effect of 5-HT_6 agonists on *in vivo* neurotrasmission (Schechter *et al.*, 2008). The agonists utilized were WAY-181187 and WAY-208466, two *N*-aryl-sulphonyl-tryptamine derivatives, with high affinity (low nanomolar range) and an over two orders of magnitude higher affinity for 5-HT_6 receptors as compared to other 5-HT receptors (Schechter *et al.*, 2008). These compounds behaved as full agonists on adenylate cyclase (Schechter *et al.*, 2008) and as partial agonists in a scintillation proximity assay/antibody immunocapture procedure of coupling to GαS (Dupuis *et al.*, 2008). Subcutaneous administration of 30 mg/kg WAY-181187 increased GABA and decreased dialysate DA and 5-HT in the rat frontal cortex, effects reversed by a 20 min s.c. pretreatment with SB-271046. The GABA increase was rapid and long-lasting while the decrease of DA and the 5-HT was slow. The same agonist increased GABA and decreased DA in the striatum, this last effect being prevented by SB-271046. The agonist increased dialysate GABA also in the dorsal hippocampus and in the amygdala but neither in the thalamus, curiously enough, nor in the NAc (Schechter *et al.*, 2008).

To the study of Schechter *et al.* (2008), apply the same considerations made regarding the dialysis studies by Dawson and collaborators: high degree of variability as indicated by high standard errors and highly scattered changes in dialysate levels of transmitter over time.

Studies from our laboratory have failed to show any significant change in DA, NA, and 5-HT levels in dialysates taken from the medial prefrontal cortex, NAc shell (same coordinates utilized in the study of ST1936) after 10 mg/kg i.p. or i.v. of WAY-181187 (Valentini *et al.*, unpublished).

Recently, we have studied the effects of ST1936 (2-(5-chloro-2-methyl-1 *H*-indol-3-yl)-*N*,*N*-dimethylethanamine, synthesized by Sigma-Tau), a full agonist

Table I
BINDING AFFINITIES OF ST1936 FOR 5-HT RECEPTORS AND FOR OTHER RECEPTORS THAT SHOWED K_i OF LESS THAN 1 μM OR DISPLACEMENT OF MORE THAN 50% AT 1 μM CONCENTRATION (AFTER BORSINI AND BORDI, IN PREPARATION).

Receptor	Displacement at 1 μm (%)	K_i (nM)	Origin
*r*5-HT$_{1A}$		925	Rat hippocampus
*r*5-HT$_{1B}$	0		Rat cerebral cortex
5-HT$_{1D}$	26		Bovine caudate
*r*5-HT$_{2A}$		2025	Rat cerebral cortex
*h*5-HT$_{2B}$		245	H recombinant (CHO cells)
*h*5-HT$_{2C}$		1577	H recombinant (CHO cells)
*r*5-HT$_3$	17		NIE-115 cells
*r*5-HT$_4$	0		Guinea pig striatum
*h*5-HT$_{5a}$	0		H recombinant (HEK cells)
*h*5-HT$_6$		13	H recombinant (CHO cells)
*h*5-HT$_7$		168	H recombinant (CHO cells)
*r*5-HTT	20		Rat cerebral cortex
*r*a$_1$		390	Rat cerebral cortex
*h*a$_2$		300	H recombinant (CHO cells)
*r*a$_2$		300	Rat cerebral cortex

r: rat; *h*: human.

of 5-HT$_6$ receptors on adenylate cyclase, that binds to human and rat recombinant 5-HT$_6$ receptors with nanomolar affinity (K_i 13–48 nM). The compound binds with lower affinity to 5-HT$_7$, 5-HT$_{2B}$, α1, and α2 receptors (see Table I).

ST1936 (5–10–20 mg/kg i.p.) increased dialysate DA and NA in the PFCX in a dose-dependent manner. As shown in Fig. 1, ST1936 increased

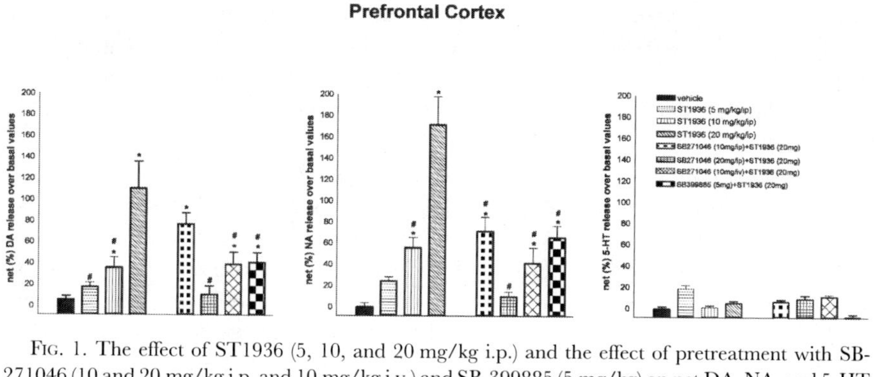

FIG. 1. The effect of ST1936 (5, 10, and 20 mg/kg i.p.) and the effect of pretreatment with SB-271046 (10 and 20 mg/kg i.p. and 10 mg/kg i.v.) and SB-399885 (5 mg/kg) on net DA, NA, and 5-HT release in the PFCX. *$p < 0.05$ versus vehicle; $^{\#}p < 0.05$ versus ST1936 (20 mg). (For interpretation of the references to color in this figure legend, the reader is referred to the Web version of this chapter.).

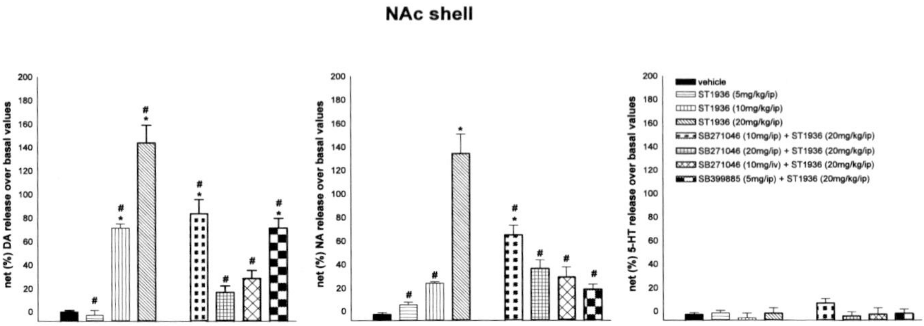

FIG. 2. The effect of ST1936 (5, 10, and 20 mg/kg i.p.) and the effect of pretreatment with SB-271046 (10 and 20 mg/kg i.p. and 10 mg/kg i.v.) or SB-399885 (5 mg/kg) on net DA, NA, and 5-HT release in the NAc shell. $*p < 0.05$ versus vehicle; $^{\#}p < 0.05$ versus ST1936 (20 mg). (For interpretation of the references to color in this figure legend, the reader is referred to the Web version of this chapter.).

dialysate DA by +58% after 10 mg/kg and by +138% after 20 mg/kg. Similarly, ST1936 increased dialysate NA by 70% and by 198% after 10 and 20 mg/kg, respectively. Pretreatment with SB-271046 (10 and 20 mg/kg i.p.), given 60 min before, dose-dependently prevented the effect of 20 mg of ST1936 on dialysate DA and NA in the medial prefrontal cortex (Fig. 3, upper panel).

ST1936 was also quite effective in raising dialysate DA and NA in the shell of the NAc (Fig. 2). Doses of 20 mg/kg i.p. increased dialysate DA and NA by 160 and 150% respectively, while 10 mg/kg raised only dialysate DA (75%). As in the prefrontal cortex, also in the NAc shell, pretreatment with SB-271046 (10 and 20 mg/kg, 1 h before), dose-dependently prevented the increase of DA and NA elicited by 20 mg/kg of ST1936 (Fig. 3, lower panel). ST1936 (5–10–20 mg/kg i. p.) was less effective in raising dialysate DA and NA in the NAc core as compared to the shell. Thus, as shown in Fig. 4, 20 mg/kg increased DA and NA by 78 and 65%, respectively, 10 mg/kg increased only DA by 55%, and 5 mg/kg was ineffective.

At all doses tested in all three investigated areas, no changes in dialysate 5-HT were observed.

These effects are unlikely to be due to blockade of the DA transporter (DAT) or of the NA transporter (NET), as ST1936 does not bind to these transporters; instead, they might be related to activation of DA and NA neuronal firing activity (Borsini *et al.*, unpublished). Agonists of 5-HT_{1A} receptors are known to increase extracellular DA in the medial prefrontal cortex (Arborelius *et al.*, 1993; Tanda *et al.*, 1994). However, these agonists also decrease extracellular 5-HT as a result of activation of somatodendritic 5-HT_{1A} autoreceptors. ST1936 does not affect extracellular 5-HT levels and this would exclude a role of 5-HT_{1A} receptors in its DA stimulant actions.

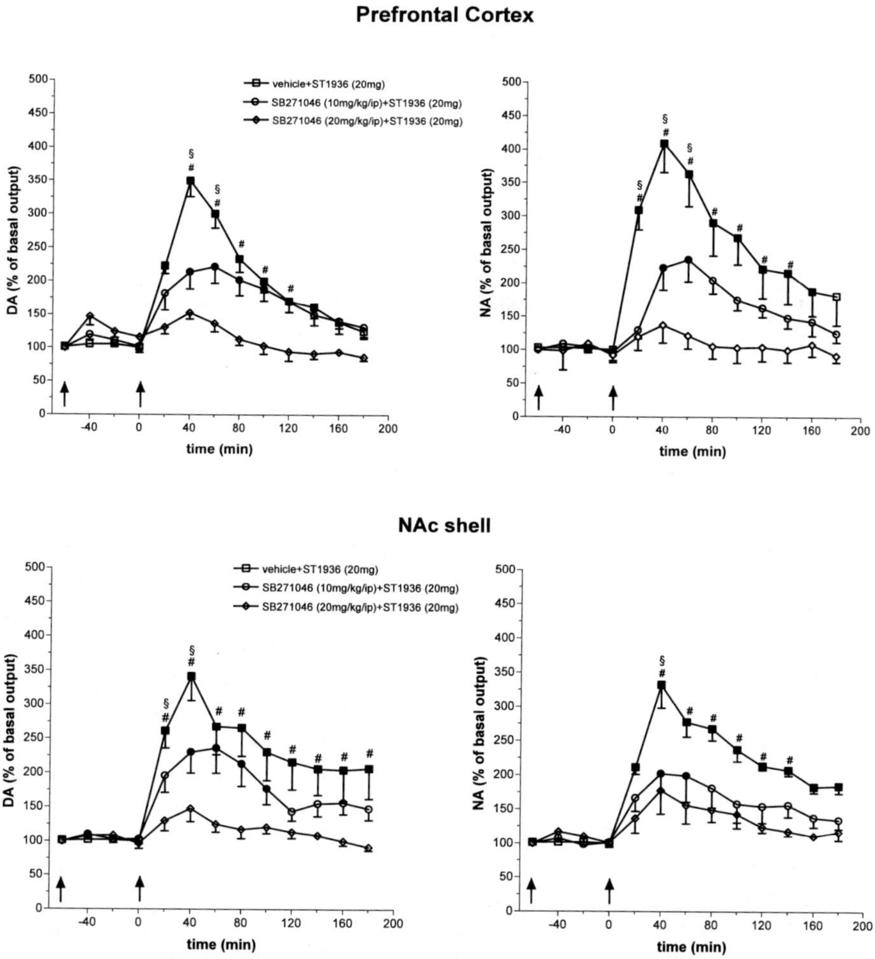

FIG. 3. Antagonism by SB-271046 on ST1936-induced DA and NA release in the PFCX and NAc shell. Filled symbol: $p < 0.05$ versus basal value; $^{\#}p < 0.05$ versus the correspondent time-point of pretreated group (SB-271046 20 mg); $^{\S}p < 0.05$ versus the correspondent time-point of pretreated group (SB-271046 10 mg); arrows indicate pretreatment time with saline or SB-271046 (10–20 mg/kg i.p.) and ST1936, respectively. (For interpretation of the references to color in this figure legend, the reader is referred to the Web version of this chapter.).

ST1936 has some affinity for $\alpha 2$ receptors ($K_i = 300$ nM) and blockade of $\alpha 2$ receptors is known to increase DA and NA in the medial prefrontal cortex (Millan *et al.*, 2000; Tanda *et al.*, 1996). Therefore, we cannot exclude that the peculiar pattern of the effects of ST1936 on monoamine release in the prefrontal cortex and in the NAc is the result of a combined stimulation of 5-HT$_6$ receptors and blockade of $\alpha 2$ receptors.

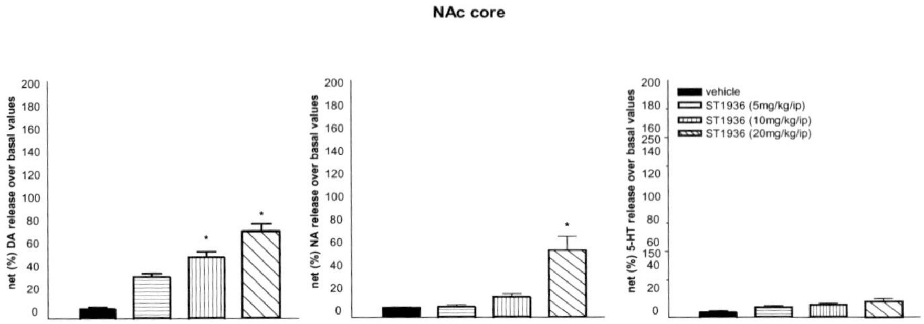

FIG. 4. The effect of ST1936 (5, 10, and 20 mg/kg i.p.) on net DA, NA, and 5-HT release in the NAc core. $*p < 0.05$ versus vehicle. (For interpretation of the references to color in this figure legend, the reader is referred to the Web version of this chapter.).

VII. Neurophysiological Effects of Drugs Acting on 5-HT$_6$ Receptors

A single study by Minabe *et al.* (2004) deals with electrophysiological effects of a 5-HT$_6$ receptor ligand. Electrophysiologically identified DA neurons of the rat substantia nigra pars compacta and VTA were recorded extracellularly and the antagonist SB-271046 was given per o.s. on single and repeated administration. The number of spontaneously active units recorded on each electrode track and their firing activity was recorded. SB-271046 (3–10 mg/kg) on single administration reduced the number of active units and the firing activity of bursting pars compacta neurons. No effects were observed in the VTA. No changes were observed on repeated administration of the antagonist. No information is available on the effect of 5-HT$_6$ receptor agonists.

West *et al.* (2009) tested the effect of the 5-HT$_6$ agonist WAY-181187 on long-term potentiation (LTP) elicited in the hippocampal CA1 field of rat brain slices. WAY-181187 reduced LTP at concentrations of 100–300 nM; at maximally effective concentration (200 nM), the reduction in slope of field EPSPs was about 80%. At the same concentration, the agonist increased spontaneous inhibitory PSPs due to the activity of GABA interneurons. All these effects were blocked by the 5-HT$_6$ antagonist SB-399885.

VIII. Role of 5-HT$_6$ Receptors in the Effect of Drugs on Neurotransmitter Function

Given the high density of 5-HT$_6$ receptors in the striatum and in particular in the ventral striatum, it was anticipated that these receptors would play a role in the effects of centrally acting drugs on striatal neurotransmitter function.

Frantz *et al.* (2002) studied the effect of the 5-HT_6 antagonist prodrug SB-258510A on the release *in vivo* of DA elicited by amphetamine in the medial prefrontal cortex and in the NAc. SB-258510A (3 mg/kg i.p.) potentiated the increase in dialysate DA induced by systemic amphetamine in the frontal cortex but not in the NAc, where only a tendency to a potentiation was observed. The antagonist also dose-dependently potentiated the locomotor activity induced by amphetamine starting from doses of 3 mg/kg but, at the same doses, it did not affect cocaine-stimulated activity. Consistent with a potentiation of the reinforcing properties of amphetamine, SB-258510A also reduced responding for amphetamine intravenous self-administration in rats trained with a dose of amphetamine on the descending limb of the dose–response curve and increased the breaking point (maximal ratio supported) in a progressive ratio paradigm. These observations indicate that blockade of 5-HT_6 receptors acutely potentiates DA-releasing amphetamine actions. The mechanism of this potentiating effect is obscure. Also puzzling is the differential effect on amphetamine as compared to cocaine.

Dawson *et al.* (2003) confirmed that an antagonist of 5-HT_6 receptors, SB-271046, at the dose of 1 and 10 mg/kg s.c. 2 h before amphetamine (0.3 mg/kg s.c.) potentiated amphetamine-induced increase of dialysate DA in the striatum, the lower dose being more effective than the higher one. This dose of amphetamine failed, by itself, to increase 5-HT in dialysates, but pretreatment with the antagonist made amphetamine effective in this respect; the two doses of antagonist were equieffective.

Local dialysis with amphetamine (100 nM) also increased DA in dialysates, but this effect was not potentiated by systemic SB-271046. No change was observed in dialysate glutamate.

Also in the study by Dawson and collaborators, the dialysis data are characterized by a high degree of variability and scatter along the time–effect curve.

IX. Transductional and IEG Effects of Systemic 5-HT_6 Receptor Activation

In contrast with the availability with a large number of specific 5-HT_6 receptor antagonists, few agonists of 5-HT_6 receptors are available and, for only three of them, some information is available on their effects on transductional effectors and immediate early genes (IEG).

Brain-derived neurotrophic factor (BDNF) has been implicated in the pathogenesis of depression and in the mechanism of action of antidepressants (for review, see Castren and Rantamaki, 2010; Kozisek *et al.*, 2008).

The 5-HT_6 agonist LY-586713 (1 mg/kg s.c.) has been reported by de Foubert *et al.* (2007) to increase BDNF mRNA expression in the rat parietal cortex 18 h after 0.1 mg/kg s.c., in the hippocampus (CA1 and dentate gyrus) after 1 mg/kg, and to be devoid of effects on the hippocampus, pyriform, and parietal cortex after 10 mg/kg. Therefore, the dose–response curve for the effects of the agonist on hippocampal BDNF was bell shaped. Pretreatment with the antagonist SB-271046 effects of the agonist resulted in a decrease of BDNF expression in the hippocampus compared to controls. Thus, combined blockade of 5-HT_6 receptors and their stimulation by LY-586713 result in effects different from those elicited by each manipulation alone, suggesting that the blockade of 5-HT_6 receptors allows the expression of some non-5-HT_6 actions of LY-586713 that are repressed by the activation of 5-HT_6 receptors.

de Foubert *et al.* (2007) also studied the effect of a single dose of LY-586713 (1 mg/kg s.c.) on the mRNA expression, estimated by autoradiography, of the IEG activity-regulated cytoskeletal-associated protein (*Arc*). The expression of this IEG increases in a region-specific manner in the rat CNS after chronic rather than acute antidepressant treatment and, similar to BDNF, has been implicated in the antidepressant mechanism of action (Pei *et al.*, 2003, 2004).

LY-586713 increased *Arc* in the hippocampus, parietal, orbital, cingulate, and parietal cortex. The 5-HT_6 antagonist SB-271046 did not affect Arc expression in the hippocampus and pyriform cortex by itself and prevented the increase induced by the agonist. However, in the cingulate and orbital cortex, the antagonist increased Arc expression and its association with the agonist did not affect the changes induced by each drug alone (de Foubert *et al.*, 2007). Therefore, also in the case of Arc, the effects of the agonist are complex and are not necessarily due to an action on 5-HT_6 receptors; thus, activation of Arc can be induced by 5-HT_2 receptor agonism (Pei *et al.*, 2003). However, since the receptor specificity and intrinsic activity of LY-586713 are unknown, it is difficult to make any precise hypothesis upon the mechanism of the effects of this agonist on BDNF and Arc.

Svenningsson *et al.* (2007) reported that fluoxetine, an antidepressant that blocks SERT, increased extracellular 5-HT, expression of c-fos mRNA and phosphorilation in Ser845 and reduced the immobility on a tail suspension test, a model of antidepressant activity. These effects were blocked by the 5-HT_6 antagonist SB-271046 and mimicked by the 5-HT_6 agonist EMDT, which in turn also reduced the immobility in the tail suspension test. However, the 5-HT_6 dependence of the biochemical effects of EMDT, that is, their sensitivity to 5-HT_6 blockade, was not tested. These observations are consistent with a role of 5-HT_6 receptors in the antidepressant properties of fluoxetine. However, the use of mice, whose 5-HT_6 receptors are, in contrast to those of the rat, so different from the human ones, makes these results not readily applicable to humans.

X. Behavioral Pharmacology of 5-HT_6 Receptors

Generation of mouse lines bearing deletions or null mutations of genes encoding for specific neurotransmitter receptors have much contributed to the knowledge of their function. This approach has been recently applied to 5-HT_6 receptors by generating a mouse line bearing a null mutation of the 5-HT_6 receptor gene (Bonasera *et al.*, 2006). The mutants did not show any overt phenotypic abnormalities in brain weight, cytoarchitecture, monoamine levels, 5-HT, DA, and NA neuron distribution, and striatal D1 and D2 receptors. No differences were observed in baseline behaviors as expressed in the open field, elevated maze, home cage activity, prepulse inhibition, Morris water maze task, motor coordination, physostigmine-induced stretching and yawning, and locomotor response to amphetamine (Bonasera *et al.*, 2006).

Mutant mice, however, showed a reduced sensitivity to the ataxic effects of 2.0 g/kg of ethanol, as indicated by a longer latency to fall from an accelerating rotarod test, and to the sedative effects of 3.5 g/kg of ethanol, as indicated by a shorter duration of loss of the righting reflex. Consistently with a reduced sensitivity to the depressant effects of ethanol, mutants showed a higher locomotor activity in response to 2 g/kg of ethanol. These differences between mutants and wild mice were pharmacodynamic rather than pharmacokinetic in nature, as indicated by the fact that no differences were observed in blood alcohol levels (Bonasera *et al.*, 2006). No differences, however, were observed in alcohol preference in a two-bottle test against water. This would suggest that the appetitive properties of ethanol were not affected by the mutation.

Failure to observe major phenotypic differences in basal brain morphology, neurochemistry, and function can be due to the action of developmental or chronic changes in neural substrates that compensate or substitute for the absent 5-HT_6 receptor function. This possibility applies in general to genetic deletion or null mutations of a given protein. However, in the case of 5-HT_6 receptors, an additional possibility is related to the fact that in the mice brain the density of these receptors is much lower than in the human brain (Hirst *et al.*, 2003), which makes the mouse not the ideal species to investigate their function in men.

In contrast with the absence of basal behavioral changes observed in mice bearing null mutations of the 5-HT receptor gene, acute or repeated manipulation of 5-HT_6 receptor function by receptor agonists and antagonists reveals a wide variety of behavioral effects in the rat.

Among the various transmitter systems that might interact and play a role in the actions of 5-HT_6 receptor-active drugs, ACh seems to be an important one. Thus, many actions of 5-HT_6 receptor-active drugs can be explained as the result of an interaction between 5-HT transmission and muscarinic ACh transmission. In some behavioral paradigms, 5-HT receptor antagonists behave as if an endogenous 5-HT$_6$ receptor tone normally acts antagonistically with ACh transmission. No interaction instead is observed with DA.

An example of the above arrangement is provided by stretching in the rat, a behavioral reaction elicited both by repeated i.c.v. infusion of antisense oligonucleotides of the 5-HT$_6$ receptor mRNA sequence (Bourson *et al.*, 1995; Sleight *et al.*, 1996) and by 5-HT$_6$ receptor antagonists (Ro 04-6790 and Ro 63-0563, Sleight *et al.*, 1998; Ro 04-6790, Bentley *et al.*, 1999; Ro 04-6790 and SB-27104, Lindner *et al.*, 2003), that is prevented by muscarinic antagonists (atropine and scopolamine), but is not affected by DA receptor antagonists such as haloperidol (Bentley *et al.*, 1999; Bourson *et al.*, 1995; Lindner *et al.*, 2003; Sleight *et al.*, 1996, 1998). The same applies to turning behavior in the unilaterally 6-OHDA-lesioned rat. Thus, 5-HT$_6$ receptor antagonists do not affect the ipsilateral turning induced by an indirect DA receptor stimulant such as amphetamine, but reduce the ipsilateral turning induced by the muscarinic antagonists atropine and scopolamine (Bourson *et al.*, 1998). The putative anticholinergic influences of 5-HT$_6$ transmission on striatal functions, however, must be quite indirect, since the 5-HT$_6$ antagonist Ro 04-6790 does not induce extrapyramidal effects and does not affect the catalepsy induced by DA receptor antagonists acting on D2 (haloperidol) and on D1 (SCH23390) receptors (Bourson *et al.*, 1998). This in turn would exclude that blockade of 5-HT$_6$ receptors contributes to the reduced extrapyramidal side effects of those atypical antipsychotics that bind with high affinity to 5-HT$_6$ receptors (see Section IV).

XI. Cognitive Effects of 5-HT$_6$ Receptor Antagonists and Agonists

The hypothesis of a functional antagonism between 5-HT$_6$ and muscarinic ACh receptors applies in particular to the role of 5-HT$_6$ transmission in cognitive functions, except that in the case of attentional and memory functions ACh plays a positive role, while in extrapyramidal functions it plays a negative role. Thus, impairment of ACh transmission in the striatum has antiparkinsonian effects while in the cortex it has procognitive effects.

A large number of studies indicate that blockade of 5-HT$_6$ receptors improves performance in various learning and memory paradigms. This field has been reviewed by Fone (2008).

In rats, acute administration of the 5-HT$_6$ receptor antagonists Ro 04-6790 (5 mg/kg i.p.), SB-357134 (1–30 mg/kg p.o.), and SB-399885 (1–30 mg/kg p.o.) improved memory consolidation in an autoshaping Pavlovian/instrumental learning paradigm (Meneses, 2001; Perez-Garcia and Meneses, 2005) and reversed scopolamine-induced impairment in a step-through passive avoidance task

(Foley *et al.*, 2004). BGC20-761, a tryptamine 5-HT_6 receptor antagonist, also reversed scopolamine-induced deficits in a social recognition paradigm and improved a time delay-induced impairment in a novel object recognition test in rats (Mitchell *et al.*, 2007). Acute administration of selective 5-HT_6 receptor antagonists and antisense oligonucleotides directed against the initiation codon of 5-HT_6 receptor mRNA improved retention of spatial learning in the Morris water maze (Foley *et al.*, 2004; Rogers and Hagan, 2001; Woolley *et al.*, 2001) and recognition memory in novel object discrimination paradigms (King *et al.*, 2004, 2009). Furthermore, the 5-HT_6 receptor antagonists, SB-399885 and SB-271046, significantly improved attention as indicated by reversal and extradimensional set-shifting performance (Hatcher *et al.*, 2005) and enhanced acquisition and retention of spatial learning in the water maze (Stean *et al.*, 2002).

Other studies, however, could not confirm some of these observations. Thus, Lindner *et al.* (2003) and Russell and Dias (2002) failed to replicate the procognitive effects of 5-HT_6 receptor antagonists in the Morris water maze and Lindner *et al.* (2003) did not confirm the reversal of scopolamine-induced deficits in a contextual fear conditioning paradigm and in the autoshaping paradigm. Moreover, Talpos *et al.* (2006) failed to observe any effect of acute parenteral administration of the 5-HT_6 receptor antagonist SB-271046 on any measure of behavior in the 5-choice serial reaction time task (5-csrtt) and in a delayed reward task, indicating that acute 5-HT_6 receptor blockade does not affect different aspects of attention and of impulsivity, namely, behavioral inhibition, as tested by the 5-crstt, and impulsive choice, as tested in the delayed reward task. It has been suggested (Talpos *et al.*, 2006) that these negative results are related to the acute versus chronic schedule of administration of the 5-HT_6 antagonist.

More recently, the therapeutic significance of 5-HT_6 receptor blockade in the management of the cognitive deficits of schizophrenia has been investigated in an animal model of repeated exposure to phencyclidine (PCP). The ability of the 5-HT_6 antagonists SB-271046 and SB-742457 as well as typical and atypical antipsychotics to reverse the cognitive impairments induced by PCP was assessed on two different tasks: attentional set-shifting and reversal learning (Idris *et al.*, 2010; Rodefer *et al.*, 2008). SB-271046 and sertindole, an atypical antipsychotic, but not the 5-HT_{2A} antagonist M100.907 nor other atypical antipsychotics such as clozapine, olanzapine, and risperidone, reversed the PCP-induced impairments in attentional set-shifting performance, a task of executive function homologous to the human Wisconsin Card Sorting task (Rodefer *et al.*, 2008) On the other hand, another 5-HT_6 antagonist, SB-742457 as well as sertindole and the 5-HT_{2A} antagonist M100.907, improved the PCP-induced deficit in reversal learning; sertindole also improved the PCP-induced deficits in episodic memory as assessed by a novel object recognition task (Idris *et al.*, 2010). The effect of sertindole in these tasks has been related to its 5-HT_6 and 5-HT_{2A} receptor blocking properties. On the basis of these observations it has been suggested that antipsychotics that block, in addition to D2 receptors, also 5-HT$_6$ and 5-HT$_{2A}$ receptors and devoid of antimuscarinic properties might exert beneficial effects on the cognitive deficits of schizophrenia (Idris *et al.*, 2010).

In contrast with the large number of studies on the cognitive effects of 5-HT$_6$ receptor antagonists, those on 5-HT$_6$ receptor agonists are limited as only three studies are available and their results are rather puzzling.

The first study to be published has been that by Loiseau *et al.* (2008) who showed that the 5-HT$_6$ receptor agonist, WAY-181187, impaired social recognition, an effect reversed by the 5-HT$_6$ receptor antagonists SB-271046 and SB-258585. The antagonists also reversed scopolamine-induced amnesia and a delay-induced impairment of social recognition. Infusion of WAY-181187 into the frontal cortex impaired social recognition. Conversely, SB-271046, infused in the same area but not in the striatum nor in the nucleus basalis magnocellularis, reversed the delay-induced impairment of recognition. These results are consistent with an amnesic role of frontal cortical 5-HT$_6$ receptors and a promnesic effect of their blockade. Results opposite to those of Loiseau *et al.* (2008) have been reported by Kendall *et al.* (2010) after administration of two different 5-HT$_6$ receptor agonists, E-6801 and EMD-386088, which produced dose-related increases in novel object recognition, an effect also induced by the antagonists SB-271046 and Ro 04-6790. Subeffective doses of the two agonists E-6801 and EMD-386088, when coadministered with SB-271046, also significantly enhanced object-recognition memory. E-6801 also reversed scopolamine-induced impairment of object recognition. The discrepancies between the studies by Loiseau *et al.* (2008) and Kendall *et al.* (2010) could be explained by a different intrinsic activity of the WAY-181187 agonist as compared to E-6801 and EMD-386088. However, a procognitive effect of WAY-181187 has been reported also by Burnham *et al.* (2010) who investigated the effect of the agonist in an extradimensional attentional set-shifting task. The effect of the drug on the expression of Fos, an IEG, in the medial prefrontal cortex was also examined. WAY-181187 facilitated extradimensional set-shifting but did not change intradimensional set-shifting and reversal. This effect was blocked by the selective 5-HT$_6$ antagonist SB-399885, which alone had no effect. WAY-181187 enhanced extradimensional set-shifting even when administered after the attentional set had been acquired, thus excluding impairments in attentional set formation. In separate experiments, at a dose that increased extradimensional set-shifting, WAY-181187 increased Fos-like immunoreactivity in the medial prefrontal cortex in a SB-399885-sensitive manner.

Therefore, the two studies by Burnham *et al.* (2010) and by Kendall *et al.* (2010) seem to agree on the ability of 5-HT$_6$ receptor agonists to produce cognitive effects superimposable to those of the antagonists. From this point of view, the observations of Kendall *et al.* (2010) are particularly striking, since the antagonists made effective a subeffective dose of the agonist. This observation is difficult to interpret. One might suggest that 5-HT$_6$ receptors in different brain areas have different inputs of endogenous 5-HT and differential influence on cognitive functions. Thus, 5-HT$_6$ receptor antagonists are likely to act on high 5-HT tone receptors while 5-HT$_6$ receptor agonists act on low 5-HT tone receptors. In turn, 5-HT$_6$ receptors might exert opposite influences on cognitive function depending on their intrinsic 5-HT tone. These differential functions, in turn, might be related to different transduction mechanisms (see Section III).

XII. Role of 5-HT$_6$ Receptors in Reward and Reinforcement

The literature available on the role of 5-HT$_6$ receptors in reward and reinforcement and on the behavioral properties of drugs of abuse is very limited, amounting, to date, to not more than seven reports. One study deals with learning of instrumental learning for food reinforcement (Mitchell *et al.*, 2007). Three studies deal with the role of 5-HT$_6$ receptors in the locomotor stimulant, rewarding and reinforcing properties of amphetamine and cocaine (Ferguson *et al.*, 2008; Frantz *et al.*, 2002; van Gaalen *et al.*, 2010), two with the stimulus properties of nicotine (Young *et al.*, 2006) and amphetamine (Pullagurla *et al.*, 2004), and one with the sedative and ataxic effects of ethanol (Bonasera *et al.*, 2006). All these studies deal with the effect of antagonists and none included testing of an agonist of 5-HT$_6$ receptors. Therefore studies on agonists are needed. Recently, our group has studied the rewarding and reinforcing properties of a 5-HT$_6$ receptor agonist.

The first study on this topic has been the one by Frantz *et al.* (2002), who reported that the 5-HT$_6$ antagonist SB-258510 (1.0, 3.0, and 10 mg/kg) dose-dependently potentiated amphetamine locomotion, reduced fixed-ratio amphetamine i.v. self-administration, and, at 3.0 mg/kg, increased the breaking point for amphetamine self-administration on a progressive ratio paradigm. The antagonist, up to 10 mg/kg, failed to affect fixed-ratio responding for i.v. cocaine self-administration and, up to 3.0 mg/kg, failed to affect cocaine-induced locomotion. These effects are indicative of a potentiation of the reinforcing and locomotor activating properties of amphetamine, most likely related to a potentiation of the DA-releasing properties of the drug (see Section VIII). These effects were unrelated to the generic psychostimulant properties of amphetamine, since cocaine effects on the same measures were not affected (Frantz *et al.*, 2002).

This in turn would exclude that the interaction with amphetamine takes place on the postsynaptic site of action of DA. The authors implicate ACh neurotransmission in these effects, but this does not explain the selectivity toward amphetamine. The antagonists, however, dose-dependently reduced cue-induced reinstatement of cocaine seeking under extinction conditions as expressed by nose-poking on the active hole of previous self-administration sessions. On the basis of these observations, it has been suggested that 5-HT$_6$ antagonists might be tested for their effects on drug craving induced by cues or contexts previously associated to drugs.

van Gaalen *et al.* (2010) have recently confirmed the failure of 5-HT$_6$ receptor antagonists (SB-271046 and Ro 04-6790) to affect direct cocaine reinforcing properties.

It should be noted, however, that Ferguson *et al.* (2008) have reported that the 5-HT$_6$ antagonist Ro-4368554 facilitates the acquisition of place preference conditioned by cocaine while viral-mediated transfer of 5-HT$_6$ receptors in the NAc shell, the area where most drugs of abuse, including cocaine, preferentially increase extracellular DA (Di Chiara and Bassareo, 2007), impairs acquisition of cocaine conditioned place preference (CPP). The same viral-mediated transfer does not affect cocaine-induced locomotion and the induction of locomotor sensitization after repeated exposure to cocaine (Ferguson *et al.*, 2008). In view of the lack of effect of the 5-HT$_6$ antagonists on cocaine self-administration, the authors tend to exclude that the manipulations of 5-HT$_6$ transmission act by changing the rewarding impact of cocaine and favor the possibility that their effects on CPP are due to an action on associative learning and, in support of this explanation, quote not only the literature on the effects of 5-HT$_6$ receptor antagonists on autoshaping (see Section X1) but also the evidence from the same laboratory (Mitchell *et al.*, 2007), that 5-HT$_6$ gene transfer into the dorsomedial CPu of the rat, but not in the central CPu, impairs the acquisition of an instrumental response for food.

In drug discrimination studies, results consistent with the observations of Frantz *et al.* (2002) were obtained by Pullagurla *et al.* (2004) who reported that in rats trained to discriminate amphetamine or cocaine from saline, pretreatment with the tryptamine 5-HT$_6$ antagonist MS-245 potentiated discrimination of the amphetamine cue while, at 1.0 mg/kg, the antagonist did not affect and, at 2.0 mg/kg, it disrupted discrimination of the cocaine cue. In a later study, the same group (Young *et al.*, 2006) reported that MS-245 also potentiated the discriminative stimulus properties of nicotine (0.06 and 0.11 mg/kg s.c.) in rats trained to discriminate 0.6 mg/kg of nicotine from saline.

MS-245 has nanomolar affinity not only on 5-HT$_6$ but also on D1 receptors but is unclear whether it is an agonist or an antagonist of this subtype of DA receptors. Studies with more specific antagonists of 5-HT$_6$ receptors are be needed.

XIII. Rewarding and Reinforcing Properties of 5-HT$_6$ Agonists

No information is available on 5-HT$_6$ agonists. We have recently studied these properties on ST1936, an agonist that dose dependently increases extracellular

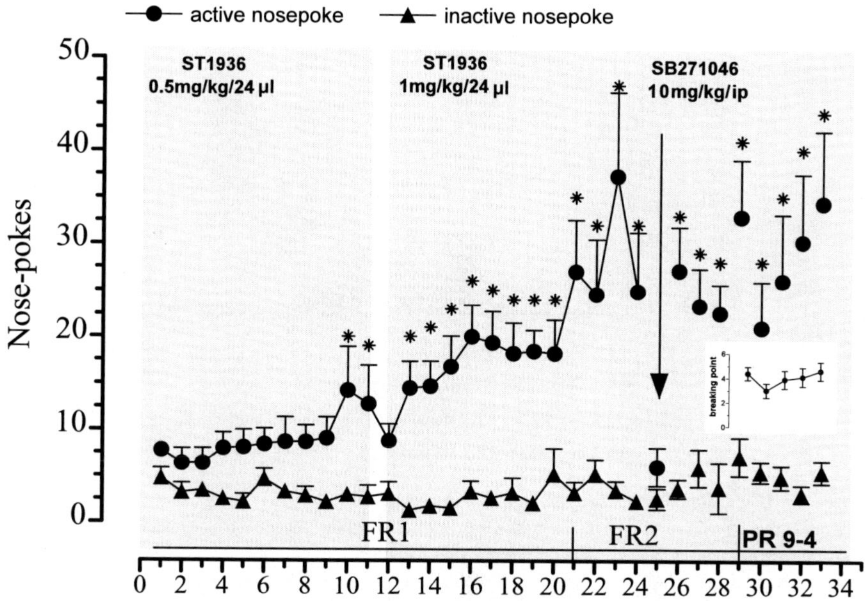

FIG. 5. Cumulative responses (nose-pokes) during ST1936 self-administration on fixed and progressive ratio, and blockade by SB-271046. Inset graph shows daily breaking point in the progressive ratio sessions. Results are expressed as mean ± SEM of cumulative nose-pokes in the active (circle) and inactive (triangle) holes during each 1-h daily SA session (1st–33rd). $*p < 0.05$ versus inactive nose-pokes. (For interpretation of the references to color in this figure legend, the reader is referred to the Web version of this chapter.).

DA and NA in the NAc shell and in the medial prefrontal cortex (Valentini *et al.*, 2010) (see Section VI.B).

ST1936, up to doses of 20 mg/kg, does not induce hypermotility. In contrast, EMD-386088 (Mattsson *et al.*, 2005), another agonist of 5-HT$_6$ receptors, elicits hypermotility at doses of 5 and 10 mg/kg i.p. (Valentini *et al.*, unpublished). Another agonist of 5-HT$_6$ receptors, WAY-181187 (Schechter *et al.*, 2008), failed to elicit hypermotility up to doses of 10 mg/kg i.p. and i.v.

As shown in Fig. 5 ST1936 is self-administered i.v. by rats at unitary doses of 0.5–1.0 mg/kg on a continuous reinforcement schedule (FR1) and on a progressive ratio schedule with breaking point of about 4.0. SB-271046 (10 mg/kg i.p.) reduced by about 80% responding for ST1936 (Valentini *et al.*, unpublished).

ST1936, at doses of 20 mg/kg i.p., also induced CPP in a two-compartment paradigm (Fig. 6) and a dose-dependent conditioned saccharin aversion (CSA) (Fig. 7) (Fenu *et al.*, unpublished).

These results show that ST1936 is provided with reinforcing properties, as it is self-administered, and with rewarding properties, since it induces CPP. The

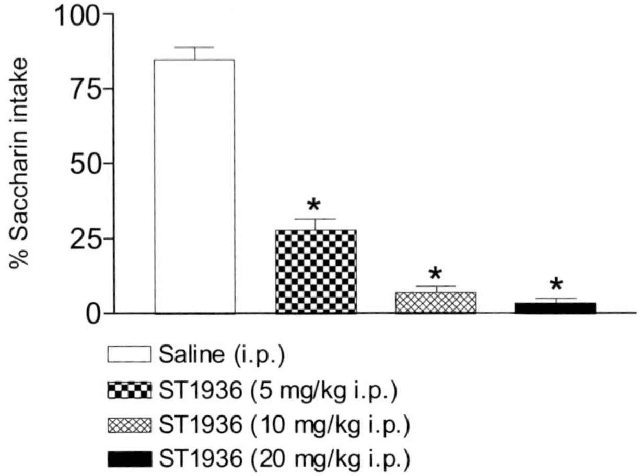

FIG. 6. Conditioned saccharin avoidance induced by ST1936 (5, 10, and 20 mg/kg i.p.) or saline administered 15 min after saccharin intake, during the conditioning phase (acquisition) of CSA. Each bar represents mean ± SEM of the percentage of saccharin intake on test day relative to the total fluid intake in a two-bottle choice with water. *$p < 0.001$ versus saline. (For interpretation of the references to color in this figure legend, the reader is referred to the Web version of this chapter.).

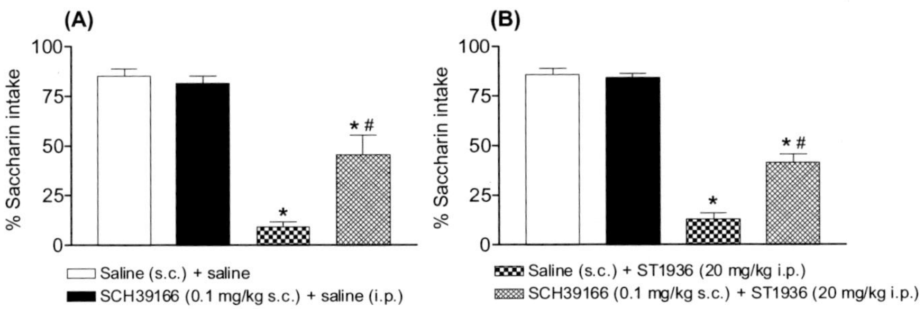

FIG. 7. Conditioned saccharin avoidance induced by ST1936 (20 mg/kg i.p.) and effect of DA D1 receptor blockade. SCH39166 (0.1 mg/kg s.c.) or saline was administered 30 min (panel A) or 45 min (panel B) after the 20-min saccharin drinking session. ST1936 or saline was given 45 min (panel A) or 60 min (panel B) after saccharin intake, during the conditioning phase (acquisition) of CSA. Each bar represents mean ± SEM of the percentage of saccharin intake on test day relative to the total fluid intake in a two-bottle choice with water. *$p < 0.001$ versus saline of the correspondent control group; #$p < 0.001$ versus saline plus ST1936. (For interpretation of the references to color in this figure legend, the reader is referred to the Web version of this chapter.).

property of inducing CSA might be an expression of the DA-releasing properties of ST1936 in the NAc shell (Fenu *et al.*, 2006, 2009, 2010). Consistent with this possibility, ST1936-induced CSA is prevented by administration of the D1 antagonist SCH39166 (Fig. 8) (Fenu *et al.*, unpublished).

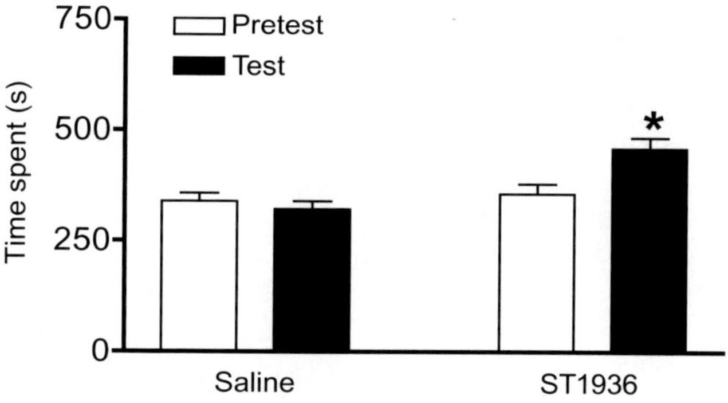

FIG. 8. Conditioned place preference induced by ST1936 (20 mg/kg i.p.) or saline after three pairing to the individually unpreferred compartment. Each bar represents mean ± SEM of the time spent (in seconds) in initially unpreferred compartment during a 15-min pretest (open bars) and a 15-min test session (closed bars). $*p < 0.001$ versus saline. (For interpretation of the references to color in this figure legend, the reader is referred to the Web version of this chapter.).

The DA releasing and mildly rewarding and reinforcing properties of 5-HT$_6$ receptor agonists might the basis for their mood-improving and antianhedonic (antidepressant?) properties (see Chapter 3).

XIV. Conclusions and Future Directions

In spite of their high density in the striatum and in particular in the ventral striatum, endogenous 5-HT$_6$ receptors do not appear to affect behaviors classically related to the ventral striatum such as motivation and reward, by a direct action in this area. Rather, the involvement of endogenous 5-HT$_6$ receptors in these functions seems to take place indirectly, via their influence on DA transmission. Thus, some (e.g., ST1936 and EMD-386088), but not all (e.g., WAY-181187), agonists of 5-HT$_6$ receptors increase extracellular DA, specifically in the shell rather than in the core of the NAc and this effect is associated with behaviors typically related to stimulation of DA transmission in this area, namely, reinforcing and rewarding properties. These differences in the behavioral effects of 5-HT$_6$ receptor agonists might be related to the fact that different agonists induce different conformations in the 5-HT$_6$ receptor with differential affinity for different transduction pathways. Thus, depending on the agonist different transduction pathways and therefore different behavioral effects can be elicited. A paradigmatic case of this kind is provided by agonists of 5-HT_{2A} receptors, whose hallucinatory and psychotomimetic properties are dependent upon the ability to induce a coupling of the receptor with Tyr kinase rather than phosphatidyl-inositol transduction pathways (Gonzalez-Maeso *et al.*, 2007). It might be just a casual circumstance but also the 5-HT_6 receptor has multiple coupling transductional pathways, including Tyr kinase and adenylate cyclase ones (see Section III).

We regard this as the frontier for future 5-HT_6 receptor research, together with the development of highly selective agonists of 5-HT_6 receptors.

References

Arborelius, L., Nomikos, G.G., Hacksell, U., and Svensson, T.H. (1993). (*R*)-8-OH-DPAT preferentially increases dopamine release in rat medial prefrontal cortex. *Acta Physiol. Scand.* **148**, 465–466.

Ballaz, S.J., Akil, H., and Watson, S.J. (2007). Analysis of 5-HT_6 and 5-HT_7 receptor gene expression in rats showing differences in novelty-seeking behavior. *Neuroscience* **147**, 428–438.

Bardo, M.T., Donohew, R.L., and Harrington, N.G. (1996). Psychobiology of novelty seeking and drug seeking behavior. *Behav. Brain Res.* **77**, 23–43.

Bentley, J.C., Bourson, A., Boess, F.G., Fone, K.C., Marsden, C.A., and Petit, N *et al.* (1999). Investigation of stretching behaviour induced by the selective 5-HT_6 receptor antagonist, Ro 04-6790, in rats. *Br. J. Pharmacol.* **126**, 1537–1542.

Boess, F.G., Monsma, F.J., Carolo, C., Meyer, V., Rudler, A., and Zwingelstein, C *et al.* (1997). Functional and radioligand binding characterization of rat 5-HT_6 receptors stably expressed in HEK293 cells. *Neuropharmacology* **36**, 713–720.

Boess, F.G., Riemer, C., Bos, M., Bentley, J., Bourson, A., and Sleight, A.J. (1998). The 5-hydroxytryptamine6 receptor-selective radioligand [3H]Ro 63-0563 labels 5-hydroxytryptamine receptor binding sites in rat and porcine striatum. *Mol. Pharmacol.* **54**, 577–583.

Bonasera, S.J., Chu, H.M., Brennan, T.J., and Tecott, L.H. (2006). A null mutation of the serotonin 6 receptor alters acute responses to ethanol. *Neuropsychopharmacology* **31**, 1801–1813.

Bourson, A., Boess, F.G., Bos, M., and Sleight, A.J. (1998). Involvement of 5-HT_6 receptors in nigrostriatal function in rodents. *Br. J. Pharmacol.* **125**, 1562–1566.

Bourson, A., Borroni, E., Austin, R.H., Monsma, F.J., and Sleight, A.J. (1995). Determination of the role of the 5-ht_6 receptor in the rat brain: a study using antisense oligonucleotides. *J. Pharmacol. Exp. Ther.* **274**, 173–180.

Brailov, I., Bancila, M., Brisorgueil, M.J., Miquel, M.C., Hamon, M., and Verge, D. (2000). Localization of 5-$HT_{(6)}$ receptors at the plasma membrane of neuronal cilia in the rat brain. *Brain Res.* **872**, 271–275.

Burnham, K.E., Baxter, M.G., Bainton, J.R., Southam, E., Dawson, L.A., and Bannerman, D.M *et al.* (2010). Activation of 5-$HT_{(6)}$ receptors facilitates attentional set shifting. *Psychopharmacology (Berl)* **208**, 13–21.

Cain, M.E., Saucier, D.A., and Bardo, M.T. (2005). Novelty seeking and drug use: contribution of an animal model. *Exp. Clin. Psychopharmacol.* **13**, 367–375.

Castren, E., and Rantamaki, T. (2010). The role of BDNF and its receptors in depression and antidepressant drug action: reactivation of developmental plasticity. *Dev. Neurobiol.* **70**, 289–297.

Dawson, L.A., Nguyen, H.Q., and Li, P. (2000). In vivo effects of the 5-HT$_{(6)}$ antagonist SB-271046 on striatal and frontal cortex extracellular concentrations of noradrenaline, dopamine, 5-HT, glutamate and aspartate. *Br. J. Pharmacol.* **130**, 23–26.

Dawson, L.A., Nguyen, H.Q., and Li, P. (2001). The 5-HT$_{(6)}$ receptor antagonist SB-271046 selectively enhances excitatory neurotransmission in the rat frontal cortex and hippocampus. *Neuropsychopharmacology* **25**, 662–668.

Dawson, L.A., Nguyen, H.Q., and Li, P. (2003). Potentiation of amphetamine-induced changes in dopamine and 5-HT by a 5-HT$_{(6)}$ receptor antagonist. *Brain Res. Bull.* **59**, 513–521.

de Foubert, G., O'Neill, M.J., and Zetterstrom, T.S. (2007). Acute onset by 5-HT$_{(6)}$-receptor activation on rat brain brain-derived neurotrophic factor and activity-regulated cytoskeletal-associated protein mRNA expression. *Neuroscience* **147**, 778–785.

Dellu, F., Piazza, P.V., Mayo, W., Le Moal, M., and Simon, H. (1996). Novelty-seeking in rats – biobehavioral characteristics and possible relationship with the sensation-seeking trait in man. *Neuropsychobiology* **34**, 136–145.

Di Chiara, G., and Bassareo, V. (2007). Reward system and addiction: what dopamine does and doesn't do. *Curr. Opin. Pharmacol.* **7**, 69–76.

Dupuis, D.S., Mannoury la, C.C., Chaput, C., Verriele, L., Lavielle, G., and Millan, M.J. (2008). Actions of novel agonists, antagonists and antipsychotic agents at recombinant rat 5-HT$_6$ receptors: a comparative study of coupling to G alpha s. *Eur. J. Pharmacol.* **588**, 170–177.

Fenu, S., Cadoni, C., and Di Chiara, G. (2010). Conditioned saccharin avoidance and sensitization to drugs of abuse. *Behav. Brain Res.* **214**, 248–253.

Fenu, S., Rivas, E., and Di Chiara, G. (2009). Differential involvement of dopamine D1 receptors in morphine- and lithium-conditioned saccharin avoidance. *Physiol. Behav.* **96**, 73–77.

Fenu, S., Spina, L., Rivas, E., Longoni, R., and Di Chiara, G. (2006). Morphine-conditioned single-trial place preference: role of nucleus accumbens shell dopamine receptors in acquisition, but not expression. *Psychopharmacology (Berl)* **187**, 143–153.

Ferguson, S.M., Mitchell, E.S., and Neumaier, J.F. (2008). Increased expression of 5-HT$_6$ receptors in the nucleus accumbens blocks the rewarding but not psychomotor activating properties of cocaine. *Biol. Psychiatry* **63**, 207–213.

Foley, A.G., Murphy, K.J., Hirst, W.D., Gallagher, H.C., Hagan, J.J., and Upton, N *et al.* (2004). The 5-HT$_{(6)}$ receptor antagonist SB-271046 reverses scopolamine-disrupted consolidation of a passive avoidance task and ameliorates spatial task deficits in aged rats. *Neuropsychopharmacology* **29**, 93–100.

Fone, K.C. (2008). An update on the role of the 5-hydroxytryptamine6 receptor in cognitive function. *Neuropharmacology* **55**, 1015–1022.

Frantz, K.J., Hansson, K.J., Stouffer, D.G., and Parsons, L.H. (2002). 5-HT$_{(6)}$ receptor antagonism potentiates the behavioral and neurochemical effects of amphetamine but not cocaine. *Neuropharmacology* **42**, 170–180.

Frederick, J.A., and Meador-Woodruff, J.H. (1999). Effects of clozapine and haloperidol on 5-HT$_6$ receptor mRNA levels in rat brain. *Schizophr. Res.* **38**, 7–12.

Gerard, C., Martres, M.P., Lefevre, K., Miquel, M.C., Verge, D., and Lanfumey, L *et al.* (1997). Immuno-localization of serotonin 5-HT$_6$ receptor-like material in the rat central nervous system. *Brain Res.* **746**, 207–219.

Glatt, C.E., Snowman, A.M., Sibley, D.R., and Snyder, S.H. (1995). Clozapine: selective labeling of sites resembling 5HT6 serotonin receptors may reflect psychoactive profile. *Mol. Med.* **1**, 398–406.

Gonzalez-Maeso, J., Weisstaub, N.V., Zhou, M., Chan, P., Ivic, L., and Ang, R *et al.* (2007). Hallucinogens recruit specific cortical 5-HT$_{(2A)}$ receptor-mediated signaling pathways to affect behavior. *Neuron* **53**, 439–452.

Hamon, M., Doucet, E., Lefevre, K., Miquel, M.C., Lanfumey, L., and Insausti, R *et al.* (1999). Antibodies and antisense oligonucleotide for probing the distribution and putative functions of central 5-HT$_6$ receptors. *Neuropsychopharmacology* **21**, 68S–76S.

Hatcher, P.D., Brown, V.J., Tait, D.S., Bate, S., Overend, P., and Hagan, J.J *et al.* (2005). 5-HT$_6$ receptor antagonists improve performance in an attentional set shifting task in rats. *Psychopharmacology (Berl)* **181**, 253–259.

Hirst, W.D., Abrahamsen, B., Blaney, F.E., Calver, A.R., Aloj, L., and Price, G.W *et al.* (2003). Differences in the central nervous system distribution and pharmacology of the mouse 5-hydroxytryptamine-6 receptor compared with rat and human receptors investigated by radioligand binding, site-directed mutagenesis, and molecular modeling. *Mol. Pharmacol.* **64**, 1295–1308.

Hirst, W.D., Minton, J.A., Bromidge, S.M., Moss, S.F., Latter, A.J., and Riley, G *et al.* (2000). Characterization of [(125)I]-SB-258585 binding to human recombinant and native 5-HT$_{(6)}$ receptors in rat, pig and human brain tissue. *Br. J. Pharmacol.* **130**, 1597–1605.

Hirst, W.D., Stean, T.O., Rogers, D.C., Sunter, D., Pugh, P., and Moss, S.F *et al.* (2006). SB-399885 is a potent, selective 5-HT$_6$ receptor antagonist with cognitive enhancing properties in aged rat water maze and novel object recognition models. *Eur. J. Pharmacol.* **553**, 109–119.

Hooks, M.S., Jones, G.H., Smith, A.D., Neill, D.B., and Justice, J.B. (1991). Response to novelty predicts the locomotor and nucleus accumbens dopamine response to cocaine. *Synapse* **9**, 121–128.

Hoyer, D., Clarke, D.E., Fozard, J.R., Hartig, P.R., Martin, G.R., and Mylecharane, E.J *et al.* (1994). International Union of Pharmacology classification of receptors for 5-hydroxytryptamine (serotonin). *Pharmacol. Rev.* **46**, 157–203.

Idris, N., Neill, J., Grayson, B., Bang-Andersen, B., Witten, L.M., and Brennum, L.T *et al.* (2010). Sertindole improves sub-chronic PCP-induced reversal learning and episodic memory deficits in rodents: involvement of 5-HT$_{(6)}$ and 5-HT$_{(2A)}$ receptor mechanisms. *Psychopharmacology (Berl)* **208**, 23–36.

Kendall, I., Slotten, H.A., Codony, X., Burgueno, J., Pauwels, P.J., and Vela, J.M *et al.* (2010). E-6801, a 5-HT$_{(6)}$ receptor agonist, improves recognition memory by combined modulation of cholinergic and glutamatergic neurotransmission in the rat. *Psychopharmacology (Berl)*

Kindlundh-Hogberg, A.M., Svenningsson, P., and Schioth, H.B. (2006). Quantitative mapping shows that serotonin rather than dopamine receptor mRNA expressions are affected after repeated intermittent administration of MDMA in rat brain. *Neuropharmacology* **51**, 838–847.

King, M.V., Sleight, A.J., Woolley, M.L., Topham, I.A., Marsden, C.A., and Fone, K.C. (2004). 5-HT$_6$ receptor antagonists reverse delay-dependent deficits in novel object discrimination by enhancing consolidation – an effect sensitive to NMDA receptor antagonism. *Neuropharmacology* **47**, 195–204.

King, M.V., Spicer, C.H., Sleight, A.J., Marsden, C.A., and Fone, K.C. (2009). Impact of regional 5-HT depletion on the cognitive enhancing effects of a typical 5-ht$_{(6)}$ receptor antagonist, Ro 04-6790, in the Novel Object Discrimination task. *Psychopharmacology (Berl)* **202**, 111–123.

Kozisek, M.E., Middlemas, D., and Bylund, D.B. (2008). Brain-derived neurotrophic factor and its receptor tropomyosin-related kinase B in the mechanism of action of antidepressant therapies. *Pharmacol. Ther.* **117**, 30–51.

Lacroix, L.P., Dawson, L.A., Hagan, J.J., and Heidbreder, C.A. (2004). 5-HT$_6$ receptor antagonist SB-271046 enhances extracellular levels of monoamines in the rat medial prefrontal cortex. *Synapse* **51**, 158–164.

Li, Z., Huang, M., Prus, A.J., Dai, J., and Meltzer, H.Y. (2007). 5-HT$_6$ receptor antagonist SB-399885 potentiates haloperidol and risperidone-induced dopamine efflux in the medial prefrontal cortex or hippocampus. *Brain Res.* **1134**, 70–78.

Lindner, M.D., Hodges Jr., D.B., Hogan, J.B., Orie, A.F., Corsa, J.A., and Barten, D.M *et al.* (2003). An assessment of the effects of serotonin 6 (5-HT$_6$) receptor antagonists in rodent models of learning. *J. Pharmacol. Exp. Ther.* **307**, 682–691.

Loiseau, F., Dekeyne, A., and Millan, M.J. (2008). Pro-cognitive effects of 5-HT$_6$ receptor antagonists in the social recognition procedure in rats: implication of the frontal cortex. *Psychopharmacology (Berl)* **196**, 93–104.

Marcos, B., Gil-Bea, F.J., Hirst, W.D., Garcia-Alloza, M., and Ramirez, M.J. (2006). Lack of localization of 5-HT$_6$ receptors on cholinergic neurons: implication of multiple neurotransmitter systems in 5-HT$_6$ receptor-mediated acetylcholine release. *Eur. J. Neurosci.* **24**, 1299–1306.

Mattsson, C., Sonesson, C., Sandahl, A., Greiner, H.E., Gassen, M., and Plaschke, J *et al.* (2005). 2-Alkyl-3-(1,2,3,6-tetrahydropyridin-4-yl)-1H-indoles as novel 5-HT$_6$ receptor agonists. *Bioorg. Med. Chem. Lett.* **15**, 4230–4234.

Meneses, A. (2001). Effects of the 5-HT$_{(6)}$ receptor antagonist Ro 04-6790 on learning consolidation. *Behav. Brain Res.* **118**, 107–110.

Millan, M.J., Lejeune, F., Gobert, A., Brocco, M., Auclair, A., and Bosc, C *et al.* (2000). S18616, a highly potent spiroimidazoline agonist at alpha(2)-adrenoceptors: II. Influence on monoaminergic transmission, motor function, and anxiety in comparison with dexmedetomidine and clonidine. *J. Pharmacol. Exp. Ther.* **295**, 1206–1222.

Minabe, Y., Shirayama, Y., Hashimoto, K., Routledge, C., Hagan, J.J., and Ashby Jr., C.R. (2004). Effect of the acute and chronic administration of the selective 5-HT$_6$ receptor antagonist SB-271046 on the activity of midbrain dopamine neurons in rats: an in vivo electrophysiological study. *Synapse* **52**, 20–28.

Mitchell, E.S., Sexton, T., and Neumaier, J.F. (2007). Increased expression of 5-HT$_6$ receptors in the rat dorsomedial striatum impairs instrumental learning. *Neuropsychopharmacology* **32**, 1520–1530.

Monsma Jr., F.J., Shen, Y., Ward, R.P., Hamblin, M.W., and Sibley, D.R. (1993). Cloning and expression of a novel serotonin receptor with high affinity for tricyclic psychotropic drugs. *Mol. Pharmacol.* **43**, 320–327.

Pei, Q., Sprakes, M., Millan, M.J., Rochat, C., and Sharp, T. (2004). The novel monoamine reuptake inhibitor and potential antidepressant, S33005, induces Arc gene expression in cerebral cortex. *Eur. J. Pharmacol.* **495**, 227–233.

Pei, Q., Zetterstrom, T.S., Sprakes, M., Tordera, R., and Sharp, T. (2003). Antidepressant drug treatment induces Arc gene expression in the rat brain. *Neuroscience* **121**, 975–982.

Perez-Garcia, G., and Meneses, A. (2005). Oral administration of the 5-HT$_6$ receptor antagonists SB-357134 and SB-399885 improves memory formation in an autoshaping learning task. *Pharmacol. Biochem. Behav.* **81**, 673–682.

Piazza, P.V., Deminiere, J.M., Le, M.M., and Simon, H. (1989). Factors that predict individual vulnerability to amphetamine self-administration. *Science* **245**, 1511–1513.

Piazza, P.V., Deroche-Gamonent, V., Rouge-Pont, F., and Le, M.M. (2000). Vertical shifts in self-administration dose-response functions predict a drug-vulnerable phenotype predisposed to addiction. *J. Neurosci.* **20**, 4226–4232.

Pullagurla, M., Bondareva, T., Young, R., and Glennon, R.A. (2004). Modulation of the stimulus effects of (+)amphetamine by the 5-HT$_6$ antagonist MS-245. *Pharmacol. Biochem. Behav.* **78**, 263–268.

Riemer, C., Borroni, E., Levet-Trafit, B., Martin, J.R., Poli, S., and Porter, R.H *et al.* (2003). Influence of the 5-HT$_6$ receptor on acetylcholine release in the cortex: pharmacological characterization of 4-(2-bromo-6-pyrrolidin-1-ylpyridine-4-sulfonyl)phenylamine, a potent and selective 5-HT$_6$ receptor antagonist. *J. Med. Chem.* **46**, 1273–1276.

Roberts, J.C., Reavill, C., East, S.Z., Harrison, P.J., Patel, S., and Routledge, C *et al.* (2002). The distribution of 5-HT$_{(6)}$ receptors in rat brain: an autoradiographic binding study using the radiolabelled 5-HT$_{(6)}$ receptor antagonist [(125)I]SB-258585. *Brain Res.* **934**, 49–57.

Rodefer, J.S., Nguyen, T.N., Karlsson, J.J., and Arnt, J. (2008). Reversal of subchronic PCP-induced deficits in attentional set shifting in rats by sertindole and a 5-HT$_6$ receptor antagonist: comparison among antipsychotics. *Neuropsychopharmacology* **33**, 2657–2666.

Rogers, D.C., and Hagan, J.J. (2001). 5-HT$_6$ receptor antagonists enhance retention of a water maze task in the rat. *Psychopharmacology (Berl)* **158**, 114–119.

Roth, B.L., Craigo, S.C., Choudhary, M.S., Uluer, A., Monsma Jr., F.J., and Shen, Y *et al.* (1994). Binding of typical and atypical antipsychotic agents to 5-hydroxytryptamine-6 and 5-hydroxytryptamine-7 receptors. *J. Pharmacol. Exp. Ther.* **268**, 1403–1410.

Ruat, M., Traiffort, E., Arrang, J.M., Tardivel-Lacombe, J., Diaz, J., and Leurs, R *et al.* (1993). A novel rat serotonin (5-HT$_6$) receptor: molecular cloning, localization and stimulation of cAMP accumulation. *Biochem. Biophys. Res. Commun.* **193**, 268–276.

Russell, M.G., and Dias, R. (2002). Memories are made of this (perhaps): a review of serotonin 5-HT$_{(6)}$ receptor ligands and their biological functions. *Curr. Top. Med. Chem.* **2**, 643–654.

Schechter, L.E., Lin, Q., Smith, D.L., Zhang, G., Shan, Q., and Platt, B *et al.* (2008). Neuropharmacological profile of novel and selective 5-HT$_6$ receptor agonists: WAY-181187 and WAY-208466. *Neuropsychopharmacology* **33**, 1323–1335.

Shirazi-Southall, S., Rodriguez, D.E., and Nomikos, G.G. (2002). Effects of typical and atypical antipsychotics and receptor selective compounds on acetylcholine efflux in the hippocampus of the rat. *Neuropsychopharmacology* **26**, 583–594.

Sleight, A.J., Boess, F.G., Bos, M., Levet-Trafit, B., Riemer, C., and Bourson, A. (1998). Characterization of Ro 04-6790 and Ro 63-0563: potent and selective antagonists at human and rat 5-HT$_6$ receptors. *Br. J. Pharmacol.* **124**, 556–562.

Sleight, A.J., Monsma Jr., F.J., Borroni, E., Austin, R.H., and Bourson, A. (1996). Effects of altered 5-ht$_6$ expression in the rat: functional studies using antisense oligonucleotides. *Behav. Brain Res.* **73**, 245–248.

Stead, J.D., Clinton, S., Neal, C., Schneider, J., Jama, A., and Miller, S *et al.* (2006). Selective breeding for divergence in novelty-seeking traits: heritability and enrichment in spontaneous anxiety-related behaviors. *Behav. Genet.* **36**, 697–712.

Stean, T.O., Hirst, W.D., Thomas, D.R., Price, G.W., Rogers, D., and Riley, G *et al.* (2002). Pharmacological profile of SB-357134: a potent, selective, brain penetrant, and orally active 5-HT$_{(6)}$ receptor antagonist. *Pharmacol. Biochem. Behav.* **71**, 645–654.

Svenningsson, P., Tzavara, E.T., Qi, H., Carruthers, R., Witkin, J.M., and Nomikos, G.G *et al.* (2007). Biochemical and behavioral evidence for antidepressant-like effects of 5-HT$_6$ receptor stimulation. *J. Neurosci.* **27**, 4201–4209.

Talpos, J.C., Wilkinson, L.S., and Robbins, T.W. (2006). A comparison of multiple 5-HT receptors in two tasks measuring impulsivity. *J. Psychopharmacol.* **20**, 47–58.

Tan, H.Y., Callicott, J.H., and Weinberger, D.R. (2009). Prefrontal cognitive systems in schizophrenia: towards human genetic brain mechanisms. *Cogn. Neuropsychiatry* **14**, 277–298.

Tanda, G., Bassareo, V., and Di Chiara, G. (1996). Mianserin markedly and selectively increases extracellular dopamine in the prefrontal cortex as compared to the nucleus accumbens of the rat. *Psychopharmacology (Berl)* **123**, 127–130.

Tanda, G., Carboni, E., Frau, R., and Di Chiara, G. (1994). Increase of extracellular dopamine in the prefrontal cortex: a trait of drugs with antidepressant potential? *Psychopharmacology (Berl)* **115**, 285–288.

Valentini, V., Frau, R., Bordi, F., Borsini, F. and Di Chiara, G. A microdialysis study of ST1936, a novel 5-HT6 receptor agonist. *Neuropharmacology* 2010 Dec 23. [Epub ahead of print]

van Gaalen, M.M., Schetters, D., Schoffelmeer, A.N., and De Vries, T.J. (2010). 5-HT$_6$ antagonism attenuates cue-induced relapse to cocaine seeking without affecting cocaine reinforcement. *Int. J. Neuropsychopharmacol.* **13**, 961–965.

Ward, R.P., and Dorsa, D.M. (1996). Colocalization of serotonin receptor subtypes 5-HT$_{2A}$, 5-HT$_{2C}$, and 5-HT$_6$ with neuropeptides in rat striatum. *J. Comp. Neurol.* **370**, 405–414.

Ward, R.P., Hamblin, M.W., Lachowicz, J.E., Hoffman, B.J., Sibley, D.R., and Dorsa, D.M. (1995). Localization of serotonin subtype 6 receptor messenger RNA in the rat brain by in situ hybridization histochemistry. *Neuroscience* **64**, 1105–1111.

West, P.J., Marcy, V.R., Marino, M.J., and Schaffhauser, H. (2009). Activation of the 5-HT$_{(6)}$ receptor attenuates long-term potentiation and facilitates GABAergic neurotransmission in rat hippocampus. *Neuroscience* **164**, 692–701.

Woolley, M.L., Bentley, J.C., Sleight, A.J., Marsden, C.A., and Fone, K.C. (2001). A role for 5-ht$_6$ receptors in retention of spatial learning in the Morris water maze. *Neuropharmacology* **41**, 210–219.

Young, R., Bondareva, T., Wesolowska, A., Young, S., and Glennon, R.A. (2006). Effect of the 5-HT$_{(6)}$ serotonin antagonist MS-245 on the actions of (−)nicotine. *Pharmacol. Biochem. Behav.* **85**, 170–177.

Yun, H.M., Baik, J.H., Kang, I., Jin, C., and Rhim, H. (2010). Physical interaction of Jab1 with human serotonin 6 G-protein-coupled receptor and their possible roles in cell survival. *J. Biol. Chem.* **285**, 10016–10029.

Yun, H.M., Kim, S., Kim, H.J., Kostenis, E., Kim, J.I., and Seong, J.Y *et al.* (2007). The novel cellular mechanism of human 5-HT$_6$ receptor through an interaction with Fyn. *J. Biol. Chem.* **282**, 5496–5505.

5-HT$_6$ RECEPTOR LIGANDS AND THEIR ANTIPSYCHOTIC POTENTIAL

Jørn Arnt and Christina Kurre Olsen

Lundbeck Research Denmark, H Lundbeck A/S, DK-2500 Valby, Denmark

Abbreviations

5-HT	5-hydroxytryptamine, serotonin
6-OHDA	6-hydroxydopamine
CAR	conditioned avoidance response
CIAS	cognitive impairment associated with schizophrenia
CNS	central nervous system
DA	dopamine
ED	extradimensional
EPS	extrapyramidal symptoms
GABA	γ-amino-butyric acid
LMA	locomotor activity
LSD	lysergic acid diethylamide
MAM	methylazoxymethanol
MATRICS	measurement and treatment research to improve cognition in schizophrenia
MK-801	dizocilpine
NCAM	neural cell adhesion molecule
NMDA	*N*-methyl-D-aspartate
NOR	novel object recognition
PCP	phencyclidine

INTERNATIONAL REVIEW OF NEUROBIOLOGY, VOL. 96
DOI: 10.1016/B978-0-12-385902-0.00006-1

PK	pharmacokinetic
PPI	prepulse inhibition
SSRI	Selective serotonin reuptake inhibitor
SubPCP	Subchronic PCP treatment
VTA	ventral tegmental area

I. Introduction

The dense expression of serotonin$_6$ (5-HT$_6$) receptors within dopaminergic terminal regions of central nervous system (CNS) (striatum and nucleus accumbens), as well as the expression in hippocampal and cortical regions, is compatible with the idea that this 5-hydroxytryptamine (5-HT) receptor subtype may be involved in manifestation of psychosis symptoms and cognitive impairments observed in patients with schizophrenia (Mitchell and Neumaier, 2005; Roberts *et al.*, 2002; Woolley *et al.*, 2004). Furthermore, some antipsychotic drugs display high affinity for 5-HT$_6$ receptors in addition to their dopamine (DA) D$_2$ antagonist effects, although 5-HT$_6$ antagonism is not a unique feature (Arnt and Skarsfeldt, 1998; Schotte *et al.*, 1996). The density of the drug target is unchanged, at least in dorsolateral prefrontal cortex, in postmortem brains from schizophrenia patients in the single available study (East *et al.*, 2002).

Schizophrenia is a heterogeneous syndrome, consisting of different symptom clusters with distinct pathophysiology, usually referred to as positive, negative, and cognitive symptom domains. The present overview attempts to summarize the evidence for involvement of 5-HT$_6$ receptors in the symptomatology of these domains, and for the therapeutic potential of 5-HT$_6$ receptor ligands. Schizophrenia is also commonly associated with affective disturbances, but since there is little knowledge of disease biology of these, which are specific for schizophrenia, this will not be described here. The reader is referred to Chapter 3 in this book.

II. Effects in Models of Positive Psychotic Symptoms and Extrapyramidal Symptoms (EPS)

A variety of animal models are available for the detection of antipsychotic efficacy and EPS, the main feature being their sensitivity to DA antagonism in the limbic and striatal DA areas, respectively (Arnt and Skarsfeldt, 1998). For simplicity models used for antipsychotic efficacy will be referred to as models of positive symptoms and include inhibitory effect on hyperactivity induced by a DA stimulant (amphetamine; cocaine or a DA agonist, e.g., apomorphine) or hyperactivity induced by an *N*-methyl-D-aspartate (NMDA) antagonist (phencyclidine (PCP) or dizocilpine (MK-801)). Reversal of other effects induced by DAergic and glutamatergic compounds are also frequently used for predicting antipsychotic activity, for example, inhibitory effects on latent inhibition and prepulse inhibition (PPI) of sensory gating. These responses are not exclusively modeling positive symptoms, as attentive and cognitive functions to some extent are involved as well (Geyer, 2008; Moser *et al.*, 2000). A classic and well-validated psychosis model is inhibition of conditioned avoidance response (CAR), in which no pharmacological tool is needed to induce disturbed behavior (Arnt, 1982; Olsen *et al.*, 2008).

In contrast, EPS are most commonly modeled by measuring the cataleptic response to an antipsychotic in rodents or, more rarely, by observing EPS features in monkeys (dystonia and bradykinesia). While all DA antagonists are fully efficacious in models of positive symptoms, several second-generation antipsychotics (with clozapine as the prototype) have additional target effects, which decrease or prevent expression of EPS in rodents and nonhuman primates (Arnt and Skarsfeldt, 1998).

Despite the high expression level of 5-HT_6 receptors in DA-rich areas of the brain, mainly associated with GABAergic neurons, there is limited evidence for 5-HT_6 agonist- or antagonist-induced effects in psychosis and EPS test models. An overview of results obtained with 5-HT_6 antagonists is shown in Table I. The 5-HT_6 antagonist Ro 04-6790 did not influence amphetamine-induced circling behavior in rats with unilateral 6-hydroxydopamine (6-OHDA)-induced lesions and did not induce catalepsy. Furthermore, the effects of a single high dose of SCH23390 (DA D_1 antagonist) and haloperidol (DA D_2 antagonist) were not modified by Ro 04-6790 (Bourson *et al.*, 1998). In parallel experiments using the 6-OHDA lesion model, scopolamine-induced circling behavior was inhibited by the 5-HT_6 antagonist, suggesting a preferential influence on cholinergic function, consistent with studies reporting 5-HT_6 antagonist-scopolamine interactions mediating "stretching behavior" (Bentley *et al.*, 1999).

In accordance with the Bourson *et al.* (1998) data, another study failed to detect effects of Ro 04-6790 and Ro 65-7199 on PPI impairments induced by apomorphine, PCP, or lysergic acid diethylamide (LSD), and did not influence expression of latent inhibition (Leng *et al.*, 2003). Some discrepant data also exist for Ro 04-6790, showing inhibition of MK-801-induced hyperactivity and stereotyped behavior (Pitsikas *et al.*, 2008). A more recently developed compound from Roche (Ro 43-68554) failed to influence PPI deficits induced by amphetamine, MK-801, scopolamine, or neonatal hippocampus lesions (Mitchell and Neumaier, 2008; Schreiber *et al.*, 2007), while PPI disruption induced by apomorphine was antagonized (Mitchell and Neumaier, 2008). It had no effect on cocaine-induced hyperactivity (Ferguson *et al.*, 2008). SB-271046 from Glaxo was also found ineffective across most models, except for a reversal of amphetamine-induced PPI

Table I

Overview of 5-HT_6 Antagonist Profiles in Models of Positive and Negative Psychotic Symptoms, Sedation, and EPS

Model	Ro 04-6790	Ro 43-68554	Ro 65-7199	SB-271046	SB-258510A	Lu AE58054	5-HT_6 KO mice
Positive symptoms							
LMA, amphetamine				NE (7)	Incr (9)	Incr (10)	NE (14)
LMA, cocaine		NE (4)			NE (9)		
LMA, MK-801	+ (1)						
LMA, isolation reared rats						NE (11)	
Circling, amphetamine	NE (2)						
Circling, scopolamine	+ (2)						
CAR						NE (12)	
VTA population activity, 3 week				NE (8)			
PPI, amphetamine	NE (3)	NE (5)		+ (7)			NE (14)
PPI, apomorphine	NE (3)	+ (6)	NE (3)				
PPI, PCP			NE (3)	NE (7)			
PPI, MK-801	NE (3)	NE (5)					
PPI, LSD			NE (3)				
PPI, scopolamine		NE (6)					
PPI, neonatal hippocampus lesion	NE (3)	NE (5)					
Latent inhibition			NE (3)				
Negative symptoms							
PCP, social interaction				NE (7)			
Subchronic PCP, social avoidance						NE (13)	
EPS and sedation							
LMA, spontaneous	NE (2)			NE (7)		NE (10)	
Catalepsy	NE (2)			NE (7)		NE (12)	

+: Significant effect (reference no. in parentheses); NE: no effect; Incr: increased or prolonged effect. References: (1) Pitsikas *et al.*, 2008; (2) Bourson *et al.*, 1998; (3) Leng *et al.*, 2003; (4) Ferguson *et al.*, 2008; (5) Schreiber *et al.*, 2007; (6) Mitchell and Neumaier, 2008; (7) Pouzet *et al.*, 2002; (8) Minabe *et al.*, 2004; (9) Frantz *et al.*, 2002; (10) Arnt, unpublished; (11) Fabricius *et al.*, unpublished; (12) this chapter; (13) Neill *et al.*, unpublished; (14) Bonasera *et al.*, 2006.

deficits (Pouzet *et al.*, 2002). Similarly, effects of SB-271046 on population DA neuron activity in ventral tegmental area (VTA) were investigated using electrophysiological methods. In this model acute and subchronic (at least for 3 weeks) treatment with antipsychotics increase or decrease, respectively, the number of active DA neurons in VTA (Bunney and Grace, 1978; Grace *et al.*, 1997; Skarsfeldt, 1992). SB-271046 had no effects on DA neuron activity (Minabe *et al.*, 2004). However, in the once-daily dosing regimen for 3 weeks it is unknown whether drug exposure and, accordingly, receptor occupancy was sufficient to ensure detection of a potential effect, as SB-271046 is known to have limited brain penetration (Upton *et al.*, 2008).

Finally, Table I summarizes published and unpublished data generated with a more recently developed 5-HT_6 antagonist, Lu AE58054 (Arnt *et al.*, 2010). The compound was ineffective across various models. It is to our knowledge, the only 5-HT_6 antagonist that has been tested in the CAR model (Figs. 1 and 2). Amphetamine hyperactivity data were inconclusive, as we discovered that Lu AE58054 increased plasma and brain amphetamine concentrations. Accordingly,

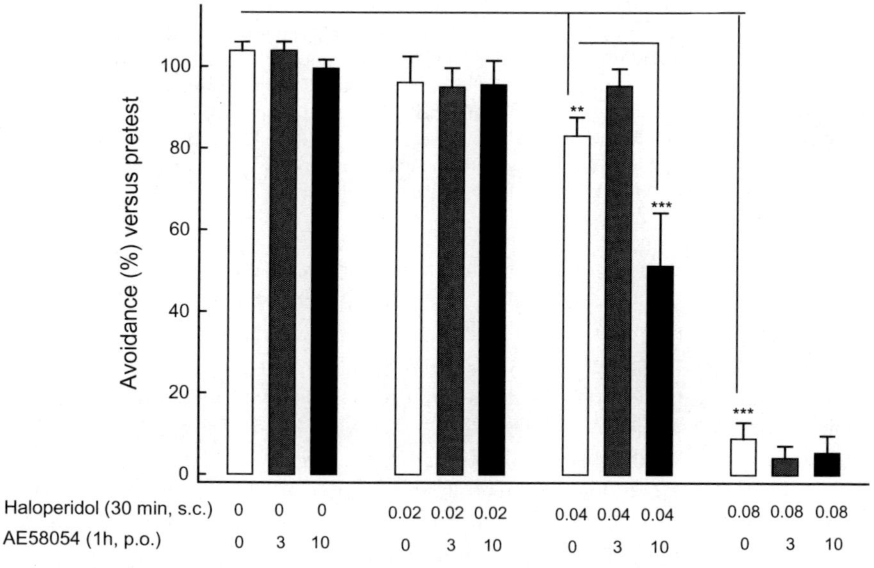

FIG. 1. Conditioned avoidance response (CAR) in rats after treatment with Lu AE58054 alone and in combination with haloperidol. Lu AE58054 (3 or 10 mg/kg, p.o.) or its vehicle (10% solution of HP-β-cyclodextrin) was administered 1 h before test, and haloperidol (0.02, 0.04, and 0.08 mg/kg, s.c.) or its vehicle (saline added minimum of tartaric acid) was injected 30 min before test. Avoidance responding was tested as described in detail (Olsen *et al.*, 2008). Ordinate indicates percentage avoidances compared with a pretest session (mean/SEM, $n = 6$–23). Lu AE58054 had no effect when given alone, but the 10 mg/kg dose enhanced the effect of haloperidol 0.04 mg/kg. $^{**}P < 0.01$, $^{***}P < 0.001$ (two-way ANOVA, followed by Student–Newman–Keuls post hoc test).

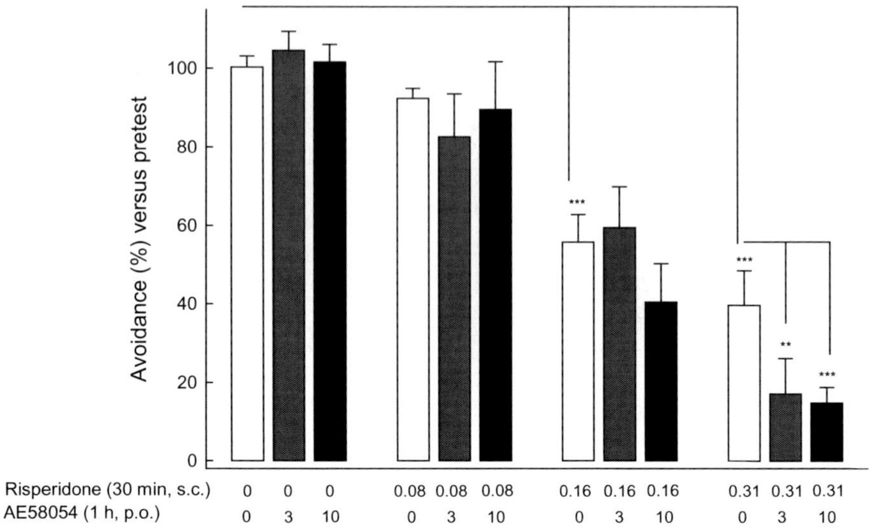

FIG. 2. Conditioned avoidance response (CAR) in rats after treatment with Lu AE58054 given alone and in combination with risperidone. Lu AE58054 (3 or 10 mg/kg, p.o.) or its vehicle (10% solution of HP-β-cyclodextrin) was administered 1 h before test, and risperidone (0.08, 0.16, and 0.31 mg/kg, s.c.) or its vehicle (saline added minimum of diluted hydrochloric acid) was injected 30 min before test. Avoidance responding was tested as described in detail (Olsen *et al.*, 2008). Ordinate indicates percentage avoidances compared with a pretest session (mean/SEM, $n = 8$–23). Lu AE58054 had no effect when given alone, but both doses enhanced the effect of risperidone 0.31 mg/kg. $^{**}P < 0.01$, $^{***}P < 0.001$ (two-way ANOVA, followed by Student–Newman–Keuls post hoc test).

the amphetamine hyperactivity response was prolonged (Arnt, unpublished observations). Such pharmacokinetic (PK) interactions are rarely controlled for in the pharmacology literature, are specific for each compound, and may thus contribute to some of the inconsistencies shown in Table I. In another model using rats reared in isolation, the rats show spontaneous DAergic hyperactivity that can be inhibited by antipsychotics (Fone and Porkess, 2008). Lu AE58054 did not affect either the hyperactivity in the isolated reared rats or the locomotor activity (LMA) in group-housed control rats, at a dose leading to high receptor occupancy (Arnt *et al.*, 2010). Risperidone preferentially reversed the hyperactivity response in rats reared in isolation, as expected (Fabricius *et al.*, unpublished data).

Similar to the paradoxical enhancing effect of Lu AE58054 on amphetamine-induced hyperactivity the same finding has been reported in another study using the 5-HT_6 antagonist SB-258510A (Frantz *et al.*, 2002). Cocaine-induced hyperactivity was in contrast not modified. Exposure levels of amphetamine are unknown in this study, but it is tempting to speculate whether the increased amphetamine response is caused by a PK interaction for this drug combination as well.

In addition to antagonist experiments, a limited data set has been obtained in mice with a genetic null mutation of the 5-HT_6 receptor gene. Again, these mice had a normal PPI and amphetamine-induced hyperactivity response (Bonasera *et al.*, 2006).

Overall, it is concluded that effects of 5-HT_6 antagonists in models of positive psychotic symptoms and EPS are limited and of questionable clinical relevance. The discrepant findings are likely caused by lack of specific 5-HT_6-mediated effects in the psychosis models, and the sporadic effects may be nonspecific for various reasons (e.g., PK interactions; additional unknown effects of the assumed selective antagonist). No clinical trials of selective 5-HT_6 antagonists in schizophrenia monotherapy are available.

Another question is whether 5-HT_6 receptor antagonism influences the consequence of an induced DA D_2 receptor blockade and whether antipsychotics modify density or function of 5-HT_6 receptors. There is no evidence for the latter: two-week treatment with a number of antipsychotics (with and without affinity for 5-HT_6 receptors) fail to change their density in striatum and frontal cortex (East *et al.*, 2002). The former question is discussed in the following section.

Although some antipsychotics have receptor profiles including high affinities for 5-HT_6 receptors, they are always associated with a variable range of additional target effects (e.g., on other 5-HT receptor subtypes and on α-adrenergic, muscarinic, and histaminergic receptors) (Arnt and Skarsfeldt, 1998; Schotte *et al.*, 1996). Accordingly, it is not possible to determine the precise contribution of 5-HT_6 antagonism to the antipsychotic profile. Specific combination experiments with selective 5-HT_6 antagonists and antipsychotics (without 5-HT_6 antagonism) are necessary to clarify this question. As mentioned earlier Ro 04-6790 did not modify catalepsy and amphetamine-induced circling behavior in 6-OHDA-lesioned rats (Bourson *et al.*, 1998). No other studies are available in the public domain, but unpublished studies of Lu AE58054 suggest that 5-HT_6 antagonism can moderately enhance the effects of haloperidol and risperidone on CAR and on risperidone in the catalepsy test. Both of these antipsychotics are devoid of 5-HT_6 antagonistic activity on their own. Figures 1 and 2 show an increased CAR-inhibitory effect of haloperidol and risperidone after combination with Lu AE58054 (3 and 10 mg/kg, p.o.), while Figs. 3 and 4 show an enhanced cataleptogenic effect of risperidone, but not haloperidol, after combination with Lu AE58054 (10 mg/kg, p.o.). In each experiment plasma exposures of both compounds were measured, ensuring that the effects were not due to PK interactions. These results suggest that 5-HT_6 antagonism can contribute to the profile of antipsychotics, although the changes are modest. Whether the differential enhancement of risperidone on CAR inhibition and catalepsy reflects larger 5-HT_6 receptor influence on the effects of antipsychotics in psychosis-relevant versus EPS-relevant models will need further substantiation. A crucial question is whether the interactions in the CAR model can be translated to enhanced

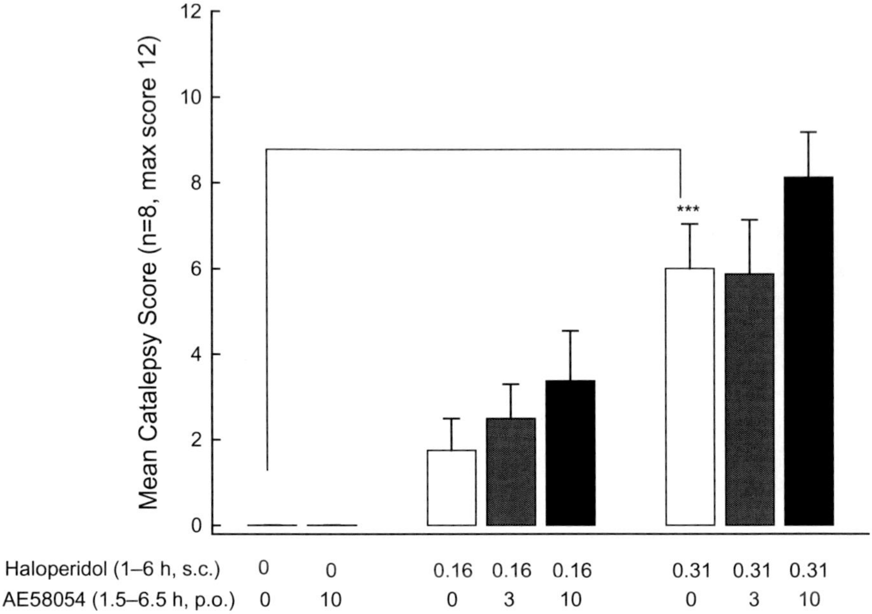

FIG. 3. Cataleptogenic effect of haloperidol in rats given alone and in combination with Lu AE58054. Lu AE58054 (3 or 10 mg/kg, p.o.) or its vehicle (10% solution of HP-β-cyclodextrin) was administered 1.5 h, and haloperidol (0.16 and 0.31 mg/kg, s.c.) or its vehicle (saline added minimum of tartaric acid) was injected 1 h before first catalepsy test. Catalepsy was evaluated on a vertical wire netting once per hour for 6 h. Rats were at each time point scored 0 (active body movements), 1 (immobile, but with head movements), or 2 (full immobility for at least 15 s). Ordinate indicates the additive score (mean/SEM, $n = 8$ per group; max score is 12). Haloperidol-induced dose-dependent catalepsy, while Lu AE58054 was ineffective, both given alone and in combination with haloperidol. $^{***}P < 0.001$ (two-way ANOVA, followed by Student–Newman–Keuls post hoc test).

antipsychotic activity or to an opportunity to reduce the minimum necessary antipsychotic dose. Limited clinical information is available, but preliminary results from a trial of Lu AE58054 added to a standard risperidone dose did not suggest improved outcome on psychosis ratings (Lundbeck internal data; for information see: http://www.lundbeck.com/investor/releases/ReleaseDetails/Release_1412413_EN.asp).

The mechanism of action of the 5-HT$_6$ and DA D$_2$ interaction is unknown, but it is speculated that 5-HT$_6$ antagonists inhibit 5-HT-mediated excitatory control of GABAergic neurons in striatum and nucleus accumbens and in this way influence the functional consequence of DA D$_2$ antagonism (Woolley *et al.*, 2004). Available evidence, based on 5-HT$_6$ receptor mRNA expression, does not suggest specific association with the direct or indirect output striatal pathway (Ward and Dorsa,

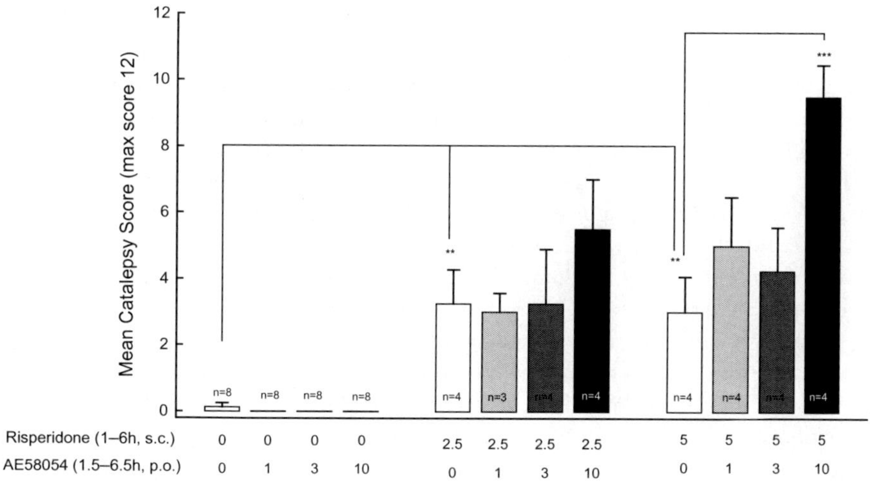

FIG. 4. Cataleptogenic effect of risperidone in rats given alone and in combination with Lu AE58054. Lu AE58054 (3 or 10 mg/kg, p.o.) or its vehicle (10% solution of HP-β-cyclodextrin) was administered 1.5 h, and risperidone (2.5 and 5.0 mg/kg, s.c.) or its vehicle (saline added minimum of dilute hydrochloric acid) was injected 1 h before first catalepsy test. Catalepsy was evaluated on a vertical wire netting once per hour for 6 h. Rats were at each time point scored 0 (active body movements), 1 (immobile, but with head movements), or 2 (full immobility for at least 15 s). Ordinate indicates the additive score (mean/SEM, $n = 8$ per group; max score is 12). Risperidone induced moderate catalepsy. Lu AE58054 was ineffective when given alone but the 10 mg/kg dose enhanced the risperidone (5 mg/kg) effect. $^{**}P < 0.01$, $^{***}P < 0.001$ (two-way ANOVA, followed by Student–Newman–Keuls post hoc test).

1996). Recent microdialysis data have shown that the 5-HT$_6$ receptor agonist WAY-181187 markedly enhance the output of γ-amino-butyric acid (GABA) in several brain regions (striatum and frontal cortex) in a pharmacologically specific manner (Mork *et al.*, 2010; Schechter *et al.*, 2008). In parallel, a moderate decrease of about 30% of extracellular DA levels was observed in striatum and frontal cortex. Paradoxically, the GABA and DA levels were unchanged in nucleus accumbens (Schechter *et al.*, 2008).

III. Effects in Models of Negative Psychotic Symptoms

The negative symptom domain presents a large unmet medical need, but remains a poorly covered area with respect to animal models and treatments. As no "golden standard" treatment is available it is difficult to validate novel models with reference compound(s). Most commonly used models with "face" (i.e., symptom) validity involve studies of social interaction deficits between animals that have been treated with acute or subchronic PCP or MK-801, or where NMDA receptor number has been genetically reduced. It has been shown that some second-generation antipsychotics can partly normalize the induced deficits (Mohn *et al.*, 1999; Sams-Dodd, 1999; Snigdha and Neill, 2008a). Only one study of a selective 5-HT_6 antagonist has been published using SB-271046, which at the doses 5 and 20 mg/kg failed to reverse the social interaction deficits induced by acute treatment with PCP (Pouzet *et al.*, 2002). In a recent yet unpublished study, acute treatment with Lu AE58054 (10 mg/kg, p.o.) similarly failed to reverse the social avoidance induced by subchronic PCP treatment for 1 week, followed by 1 week washout before conducting the experiment (Neill *et al.*, unpublished). In this model, 5-HT_{1A} partial agonists and an SSRI (fluoxetine) show significant efficacy (Snigdha and Neill, 2008b). Thus, there is presently no evidence for clinical efficacy of 5-HT_6 antagonists on negative symptom-like symptoms.

Other animal models with some association with key features of negative symptoms involve measuring effects on reinforcement and anhedonia. The reader is referred to relevant chapters in this book for further discussion.

IV. Effects in Models of CIAS (Cognitive Impairment Associated with Schizophrenia)

The involvement of 5-HT_6 receptors in cognitive function is the most well-established effect documented for this target and is the subject of some chapters in this book. When focusing on cognitive responses in animal models *relevant for the impairment in schizophrenia*, the number of preclinical (and clinical) studies of 5-HT_6 antagonists is limited.

In the last 10–15 years there has been an increased focus on the cognitive symptom domains in schizophrenia, as the integrity of these were found to have the largest impact on functional outcome (Green, 2007). The measurement and treatment research to improve cognition in schizophrenia (MATRICS) initiative in USA (http://www.matrics.ucla.edu) was established as a shared effort between academia, government, and industry to optimize and harmonize preclinical and clinical measures of cognition, as well as to identify and validate the most promising drug targets. This has enhanced the interest for CIAS as a possible separate indication for new drugs as adjunct treatments to antipsychotics. Accordingly, the activity level from both preclinical and clinical side of drug discovery and development has markedly increased, in order to improve translational (= predictive) value of models and tests to identify novel treatments of CIAS. The MATRICS initiative has classified cognitive measures relevant for CIAS into seven classes:

working memory, attention/vigilance and preattentive processing, verbal learning and memory, visual learning and memory, speed of processing, reasoning and problem solving (including executive function), and finally social cognition. Each class has its individual phenomenological and neurobiological characteristics, but all are more or less impaired in most schizophrenia patients.

For the purpose of reviewing the literature on CIAS animal models it is important to distinguish between animal *models* and *tests* to avoid confusion. The *model* is associated with the manipulation of the animal, in order to mimic pathophysiological mechanisms of CIAS, although it is acknowledged that no model possibly can replicate all aspects of this heterogeneous symptom domain. The *models* used involve either selective or combined pharmacological (e.g., acute or subchronic treatment of adult animals with PCP or other NMDA antagonists), neurodevelopmental (e.g., perinatal treatment with PCP; prenatal methylazoxymethanol (MAM) treatment of pregnant females; neonatal hippocampal lesions), environmental (e.g., isolation rearing after weaning of pups), or genetic manipulation (e.g., NMDA receptor knockdown; striatal DA D_2 receptor overexpression; modification of risk genes for schizophrenia) (Broberg *et al.*, 2009; Desbonnet *et al.*, 2009; Featherstone *et al.*, 2007; Fone and Porkess, 2008; Jentsch and Roth, 1999; Kellendonk *et al.*, 2009; Kirby *et al.*, 2009; Mohn *et al.*, 1999; Rodefer *et al.*, 2008; Tseng *et al.*, 2009). In contrast, the *test* is the cognitive response measure, for example, spatial learning in a water maze, working memory in a maze or in an operant chamber, executive function as measured by attentional set-shifting, attention performance measured in a 5-choice box, and so on. The specific combination of *model* and *test* determines the relevance of a specific study for the indication or symptom dimension in question. In the following, the available preclinical studies will be reviewed.

The literature on effect of 5-HT_6 antagonists in models of CIAS is scarce, and data on recently developed 5-HT_6 agonists in relevant models are not yet available (Table II). The table shows the few studies in cognitive tests, using induced deficits in rats with relevance for schizophrenia. Preclinical studies of impaired cognition relevant for aging and dementia (e.g., induced by scopolamine or by using aged rats) will be discussed in Chapter 7.

Ro 04-6790 improves delayed novel object recognition (NOR), and this effect is reversed by combination with MK-801 (King *et al.*, 2004). The study is, however, more relevant in the attempt to elucidate glutamatergic mechanisms involved in the 5-HT_6 antagonist-mediated improvement of memory retention than for validation of CIAS indication. But consistent with the result, another study of the same drug combination showed that Ro 04-6790 prevented the impairment of NOR induced by acute treatment with MK-801 (Pitsikas *et al.*, 2008).

Most studies in CIAS models have been performed using the so-called subchronic PCP model. The dose regimen used in the model is most often treatment with PCP for 1 week, followed by at least 1 week of withdrawal before conducting

Table II

OVERVIEW OF 5-HT_6 ANTAGONIST AND AGONIST PROFILES IN RAT MODELS OF CIAS

Model (Model, Test)	Ro 04-6790	SB-271046	SB-399885	SB-258585	SB-742457	Lu AE58054	WAY-81187[a]	E-6801[a]
MK-801, NOR	+ (1, 2)							
SubPCP, attentional set-shifting		+ (3, 4)	+ (4)					
SubPCP, NOR						+ (10)		
SubPCP, reversal learning					+ (9)	+ (11)		
Naïve rats, attentional set-shifting[b]		NE (4), (+) (5)	+ (4, 5) NE (8)				+ (8)	
Naïve rats, NOR[b]	+ (2)	+ (6)						+ (6)
Naïve rats, social cognition deficit induced by 5-HT_6 agonist[b]		+ (7)		+ (11)				

NE: No effect; +: significant reversal of deficit. References: (1) Pitsikas *et al.*, 2008; (2) King *et al.*, 2004; (3) Rodefer *et al.*, 2008; (4) Wünsch *et al.*, 2006; (5) Hatcher *et al.*, 2005; (6) Kendall *et al.*, 2010; (7) Loiseau *et al.*, 2008; (8) Burnham *et al.*, 2010; (9) Idris *et al.*, 2010; (10) Arnt *et al.*, 2010; (11) Neill *et al.*, unpublished.

[a] 5-HT_6 agonist.

[b] General cognitive test – not CIAS specific.

the cognitive testing after acute or repeated treatment with the potential CIAS drug candidate. With this dose regimen rats develop some long-term changes similar to those seen in brains of schizophrenia patients, including a decrease in expression of parvalbumin in frontal cortex and hippocampus (Abdul-Monim *et al.*, 2007; Reynolds *et al.*, 2004). Parvalbumin is a calcium-binding protein expressed in a population of GABAergic interneurons in these brain areas, and is important for proper functioning of excitatory-inhibitory networks (Lewis and Moghaddam, 2006; Lewis *et al.*, 2008). Similar changes are seen in adult rats treated with PCP in early life (Wang *et al.*, 2008) and in offsprings from rats treated during gestation (day E17) with the antimitotic compound, MAM acetate (Lodge and Grace, 2009; Lodge *et al.*, 2009; Penschuck *et al.*, 2006).

Subchronic PCP, neonatal PCP, and MAM consistently induce long-term impairments of cognitive performance in some cognitive tests relevant for deficits seen in schizophrenia patients, such as attentional set-shifting (test of executive function), NOR with short intertrial responses (test of episodic memory), and reversal learning (test of problem solving and reasoning) (Birrell and Brown, 2000; Broberg *et al.*, 2009; Egerton *et al.*, 2008; Featherstone *et al.*, 2007; Goetghebeur and Dias, 2009; Idris *et al.*, 2010; Rodefer *et al.*, 2005).

Only few selective 5-HT$_6$ antagonists have been studied in CIAS models: in the attentional set-shifting task acute treatment with SB-271046 and SB-399885 reversed the subchronic PCP-induced extradimensional (ED) shift deficits, while the latter also improved reversal learning and ED shifts in the subchronic vehicle-treated rats as well (Rodefer *et al.*, 2008; Wünsch *et al.*, 2006). Repeated treatment with SB-399885 and SB-271046 have also been tested in drug-naïve rats and showed improved performance of reversal of acquisition of compound discrimination as well as ED shifting (Hatcher *et al.*, 2005). In contrast, a recent study has shown that a full 5-HT$_6$ agonist (WAY-181817) paradoxically can improve ED shift learning in drug-naïve rats, while the 5-HT$_6$ antagonist SB-399885 was ineffective, except for preventing the agonist effect (Burnham *et al.*, 2010). These results suggest that 5-HT$_6$ receptor-mediated control of cognitive performance is more complex than originally believed. The apparent discrepancies may be due to several factors: it may implicate presence of 5-HT$_6$ receptors with opposite functional outcome at different sites in the neuronal network regulating cognition. As such, there is no evidence for molecular heterogeneity of 5-HT$_6$ receptors. Alternatively, the discrepancy may depend on state-dependent 5-HT tonus at 5-HT$_6$ receptor-expressing synapses. There is a need for tonic activity in order to reveal an effect of an antagonist, except if the receptor shows constitutive activity. The latter has not been demonstrated *in vivo* for 5-HT$_6$ receptors. In contrast, the effect of an agonist would theoretically be easier to demonstrate in conditions of low or absent tonic activity. So far, the efficacy of a 5-HT$_6$ agonist has only been demonstrated in normal rats with intact cognitive performance, while antagonists have shown consistent effects both in normal rats (with few exceptions) and in rats with impaired performance, for example, after subchronic PCP treatment.

Results with 5-HT_6 antagonists in the subchronic PCP model are available from two other tests: in the NOR test of episodic memory, Lu AE58054 has been shown to reverse PCP impairment (Table II) (Arnt *et al.*, 2010). In an operant reversal learning test rats previously treated with subchronic PCP fail to perform correctly in the reversal part of the test session, without showing deficits in the reference memory part of the test (Abdul-Monim *et al.*, 2006). SB-742457 reversed the deficit in a recent study applying the operant reversal learning task, as did the 5-HT_{2A} antagonist M100907 (Idris *et al.*, 2010). In addition, Lu AE58054 has shown full reversal in this model (Table II) (Neill *et al.*, unpublished data). The effects of Lu AE58054 in the subchronic PCP-impaired rats were observed at high 5-HT_6 receptor occupancies (>65%), when plasma exposures were compared between animals subject to cognition testing and *in vivo* receptor binding (Arnt *et al.*, 2010).

These models have been widely used for evaluation of various antipsychotics (with and without 5-HT_6 receptor affinity), and in general second-generation antipsychotics have shown similar efficacies (though with variable effective dose windows limited by response inhibition at higher doses). Two studies have found a differentiation within second-generation antipsychotics, where sertindole, but not risperidone, olanzapine, and clozapine, had efficacy in the attentional set-shifting test. In contrast, first-generation antipsychotics (haloperidol, chlorpromazine) lack efficacy (Abdul-Monim *et al.*, 2003, 2006; Goetghebeur and Dias, 2009; Grayson *et al.*, 2007; Idris *et al.*, 2010; Rodefer *et al.*, 2008). It can therefore be argued that the outcome of these tests may lack predictive value and emphasize the need for improved models and tests, since several second-generation antipsychotics are considered to have limited or almost no efficacy on cognitive impairment in schizophrenia patients (Keefe *et al.*, 2007). However, it must be remembered that the preclinical studies are performed using animals with well-defined impairments, while the patient populations used in clinical trials most often are chronic schizophrenics. These comprise a much more heterogeneous symptom group, and may show progressive decreases in cognitive function, less amenable for improvement by drugs.

Since, as earlier mentioned in the section on psychotic symptoms, some second-generation antipsychotics have high 5-HT_6 receptor affinities (Arnt and Skarsfeldt, 1998; Bymaster *et al.*, 2001; Schotte *et al.*, 1996), it may contradict the promise of developing selective 5-HT_6 antagonists for adjunct treatment of CIAS. However, the 5-HT_6-active antipsychotics, in particular clozapine and olanzapine, have several additional target effects that may mask the consequences of 5-HT_6 antagonism, such as antimuscarinic and H_1 antihistaminergic activity (Schotte *et al.*, 1996; Zhang and Bymaster, 1999). If this is correct, 5-HT_6 antagonists will not be relevant add-on treatment to all antipsychotics. Particularly those with antimuscarinic effects should be avoided, since this effect opposes 5-HT_6

actions. For other antipsychotics, add-on treatment with a selective 5-HT$_6$ antagonist has potential, as it enables the clinician to optimize the balance of 5-HT$_6$ effect and DA D$_2$ antagonism to the benefit of the individual patient.

A neglected point in preclinical validation of potential CIAS treatments is that most compounds will be used as add-on treatments with antipsychotics, since they lack efficacy on psychotic positive symptoms. Almost exclusively, however, validation of new compounds in CIAS models has been investigated using monotherapy. A major problem is that antipsychotics may confound the efficacy evaluation of a CIAS compound, since they can inhibit learning and/or memory as well as decrease response rate in cognitive tests that often are dependent on a motor response. This has, for example, been shown with several antipsychotics for spatial learning in the Morris water maze (Skarsfeldt, 1996). It should be noted that the inhibitory effects of antipsychotics on cognition may not only be due to too potent DA D$_2$ antagonism but also depend on additional effects built into antipsychotics (e.g., antimuscarinic and H$_1$ antihistaminic activity), leading to reduced motivation, learning ability, and/or sedative effects. Even though the second-generation antipsychotic olanzapine can improve performance in subchronic PCP-treated rats there is only a small efficacy window. Increasing dose levels subsequently markedly impair response rate and accuracy in the test (Abdul-Monim *et al.*, 2006). Accordingly, it will be a challenge to demonstrate ability to reverse impaired cognition in a CIAS model in combination with antipsychotics. A further issue in this respect is the relatively small efficacy window generally presented in most of the animal models of CIAS, making it statistically difficult to demonstrate added value of a CIAS drug candidate to a second-generation antipsychotic if the latter already has some efficacy in the model. No add-on studies of 5-HT$_6$ antagonists with antipsychotics have been published so far, but unpublished results (Neill *et al.*, unpublished) with Lu AE58054 suggest that its efficacy in NOR and reversal learning is maintained in the presence of a moderate dose of haloperidol (0.05 mg/kg) that produces sufficient DA D$_2$ receptor occupancy to induce functional relevant effects on DAergic activity (Arnt, 1995; Idris *et al.*, 2005).

Finally, most animal studies of CIAS treatments have used acute drug treatment, which contrasts sharply to the long-treatment duration (i.e., several months) in clinical trials. No animal studies of 5-HT$_6$ antagonists using repeated drug treatment have, to our knowledge, been performed, and only a few studies of repeated dosing with antipsychotics are available using the attentional set-shifting test when treatment is started *after* the PCP-induced impairment (McLean *et al.*, 2008; Rodefer *et al.*, 2008). In the Rodefer *et al.* study, acute and 3 weeks' treatment with either olanzapine or sertindole was compared in the standard subchronic PCP model. No differences between acute and subchronic treatment were observed: sertindole significantly reversed impairment, while olanzapine produced a nonsignificant reversal. In the McLean *et al.* study risperidone and clozapine improved attentional set-shifting performance of subchronic PCP-impaired rats after 1-week treatment, while acute treatment was not studied. The apparent similarity of acute versus repeated dose efficacy of antipsychotics on attentional set-shifting performance suggests that both effects are symptomatic rather than depend on a change in neuroplasticity. This could be mediated through acute changes in neurotransmitter availability or function, particularly of DA, GABA, acetylcholine, and glutamate in hippocampus and cortex (Dawson *et al.*, 2001; Lacroix *et al.*, 2004; Li *et al.*, 2007; Mork *et al.*, 2009; Schechter *et al.*, 2008) - effects, which have not yet been studied in disease-relevant CIAS models. We also lack studies exploring whether the reversal of cognitive impairment is associated with parallel long-term changes in disease-related marker(s), for example, expression of parvalbumin. It has been shown that subchronic treatment with 5-HT_6 antagonists can increase polysialylation of rat neural cell adhesion molecule (NCAM) in hippocampus and cortex, a mechanism thought to be involved in memory consolidation (Foley *et al.*, 2008). This result suggests that 5-HT_6 antagonists can indeed induce long-term CNS changes in drug-naïve rats, but the phenomenon needs to be studied in CIAS-relevant models as well.

V. Summary and Conclusions

The available information does not support that 5-HT_6 antagonists have significant potential for treatment of psychotic symptoms of schizophrenia as monotherapy. However, the robust efficacy demonstrated by 5-HT_6 antagonists for cognitive improvement in general, and more specifically in CIAS models, appears promising. Thus, well-controlled clinical trials are awaited with high interest. There are still many uncertainties regarding the molecular mechanisms of action, as most studies addressing this question have been made in animals without impaired behavior. Animals with induced deficits may have very different neurotransmitter dynamics compared with the normal state, which can be involved in the contradictory similar findings with antagonists and agonists in some studies.

Acknowledgments

We acknowledge the contributions by the Lu AE58054 Project Team and colleagues at Lundbeck Research, Denmark, as well as the fruitful collaboration with the group of Professor Jo Neill, University of Bradford, UK.

References

Abdul-Monim, Z., Neill, J.C., and Reynolds, G.P. (2007). Sub-chronic psychotomimetic phencyclidine induces deficits in reversal learning and alterations in parvalbumin-immunoreactive expression in the rat. *J. Psychopharmacol.* **21**, 198–205.

Abdul-Monim, Z., Reynolds, G.P., and Neill, J.C. (2003). The atypical antipsychotic ziprasidone, but not haloperidol, improves phencyclidine-induced cognitive deficits in a reversal learning task in the rat. *J. Psychopharmacol.* **17**, 57–65.

Abdul-Monim, Z., Reynolds, G.P., and Neill, J.C. (2006). The effect of atypical and classical antipsychotics on sub-chronic PCP-induced cognitive deficits in a reversal-learning paradigm. *Behav. Brain Res.* **169**, 263–273.

Arnt, J. (1982). Pharmacological specificity of conditioned avoidance response inhibition in rats: inhibition by neuroleptics and correlation to dopamine receptor blockade. *Acta Pharmacol. Toxicol.* **51**, 321–329.

Arnt, J. (1995). Differential effects of classical and newer antipsychotics on the hypermotility induced by two dose levels of D-amphetamine. *Eur. J. Pharmacol.* **283**, 55–62.

Arnt, J., Bang-Andersen, B., Bymaster, F.P., Cohen, M.P., DeLapp, N.W., Giethlen, B., Grayson, B., Kreilgaard, M., McKinzie, D.L., Neill, J., Nielsen, S.M., Nelson, D.L., Poulsen, M.N., Schauss, J. M., and Witten, L.M. (2010). Lu AE58054, a 5-HT_6 antagonist, reverses cognitive impairment induced by subchronic phencyclidine treatment in a novel object recognition test in rats. *Int. J. Neuropsychopharmacol* **13**, 1021–1033.

Arnt, J., and Skarsfeldt, T. (1998). Do novel antipsychotics have similar pharmacological characteristics? A review of the evidence. *Neuropsychopharmacology* **18**, 63–101.

Bentley, J.C., Bourson, A., Boess, F.G., Fone, K.C., Marsden, C.A., Petit, N., and Sleight, A.J. (1999). Investigation of stretching behaviour induced by the selective 5-HT_6 receptor antagonist, Ro 04-6790, in rats. *Br. J. Pharmacol.* **126**, 1537–1542.

Birrell, J.M., and Brown, V.J. (2000). Medial frontal cortex mediates perceptual attentional set shifting in the rat. *J. Neurosci.* **20**, 4320–4324.

Bonasera, S.J., Chu, H.M., Brennan, T.J., and Tecott, L.H. (2006). A null mutation of the serotonin 6 receptor alters acute responses to ethanol. *Neuropsychopharmacology* **31**, 1801–1813.

Bourson, A., Boess, F.G., Bos, M., and Sleight, A.J. (1998). Involvement of 5-HT_6 receptors in nigro-striatal function in rodents. *Br. J. Pharmacol.* **125**, 1562–1566.

Broberg, B.V., Glenthoj, B.Y., Dias, R., Larsen, D.B., and Olsen, C.K. (2009). Reversal of cognitive deficits by an ampakine (CX516) and sertindole in two animal models of schizophrenia–sub-chronic and early postnatal PCP treatment in attentional set-shifting. *Psychopharmacology (Berl)* **206**, 631–640.

Bunney, B.S., and Grace, A.A. (1978). Acute and chronic haloperidol treatment: comparison of effects on nigral dopaminergic cell activity. *Life Sci.* **23**, 1715–1727.

Burnham, K.E., Baxter, M.G., Bainton, J.R., Southam, E., Dawson, L.A., Bannerman, D.M., and Sharp, T. (2010). Activation of $5\text{-HT}_{(6)}$ receptors facilitates attentional set shifting. *Psychopharmacology (Berl)* **208**, 13–21.

Bymaster, F.P., Falcone, J.F., Bauzon, D., Kennedy, J.S., Schenck, K., DeLapp, N.W., and Cohen, M. L. (2001). Potent antagonism of 5-HT(3) and $5\text{-HT}_{(6)}$ receptors by olanzapine. *Eur. J. Pharmacol.* **430**, 341–349.

Dawson, L.A., Nguyen, H.Q., and Li, P. (2001). The $5\text{-HT}_{(6)}$ receptor antagonist SB-271046 selectively enhances excitatory neurotransmission in the rat frontal cortex and hippocampus. *Neuropsychopharmacology* **25**, 662–668.

Desbonnet, L., Waddington, J.L., and Tuathaigh, C.M. (2009). Mice mutant for genes associated with schizophrenia: common phenotype or distinct endophenotypes? *Behav. Brain Res.* **204**, 258–273.

East, S.Z., Burnet, P.W., Leslie, R.A., Roberts, J.C., and Harrison, P.J. (2002). 5-HT_6 receptor binding sites in schizophrenia and following antipsychotic drug administration: autoradiographic studies with [^{125}I]SB-258585. *Synapse* **45**, 191–199.

Egerton, A., Reid, L., McGregor, S., Cochran, S.M., Morris, B.J., and Pratt, J.A. (2008). Subchronic and chronic PCP treatment produces temporally distinct deficits in attentional set shifting and prepulse inhibition in rats. *Psychopharmacology (Berl)* **198**, 37–49.

Featherstone, R.E., Rizos, Z., Nobrega, J.N., Kapur, S., and Fletcher, P.J. (2007). Gestational methylazoxymethanol acetate treatment impairs select cognitive functions: parallels to schizophrenia. *Neuropsychopharmacology* **32**, 483–492.

Ferguson, S.M., Mitchell, E.S., and Neumaier, J.F. (2008). Increased expression of 5-HT_6 receptors in the nucleus accumbens blocks the rewarding but not psychomotor activating properties of cocaine. *Biol. Psychiatry* **63**, 207–213.

Foley, A.G., Hirst, W.D., Gallagher, H.C., Barry, C., Hagan, J.J., Upton, N., Walsh, F.S., Hunter, A.J., and Regan, C.M. (2008). The selective 5-HT_6 receptor antagonists SB-271046 and SB-399885 potentiate NCAM PSA immunolabeling of dentate granule cells, but not neurogenesis, in the hippocampal formation of mature Wistar rats. *Neuropharmacology* **54**, 1166–1174.

Fone, K.C., and Porkess, M.V. (2008). Behavioural and neurochemical effects of post-weaning social isolation in rodents-relevance to developmental neuropsychiatric disorders. *Neurosci. Biobehav. Rev.* **32**, 1087–1102.

Frantz, K.J., Hansson, K.J., Stouffer, D.G., and Parsons, L.H. (2002). 5-$HT_{(6)}$ receptor antagonism potentiates the behavioral and neurochemical effects of amphetamine but not cocaine. *Neuropharmacology* **42**, 170–180.

Geyer, M.A. (2008). Developing translational animal models for symptoms of schizophrenia or bipolar mania. *Neurotox. Res.* **14**, 71–78.

Goetghebeur, P., and Dias, R. (2009). Comparison of haloperidol, risperidone, sertindole, and modafinil to reverse an attentional set-shifting impairment following subchronic PCP administration in the rat – a back translational study. *Psychopharmacology (Berl)* **202**, 287–293.

Grace, A.A., Bunney, B.S., Moore, H., and Todd, C.L. (1997). Dopamine-cell depolarization block as a model for the therapeutic actions of antipsychotic drugs. *Trends Neurosci.* **20**, 31–37.

Grayson, B., Idris, N.F., and Neill, J.C. (2007). Atypical antipsychotics attenuate a sub-chronic pcp-induced cognitive deficit in the novel object recognition task in the rat. *Behav. Brain Res.* **184**, 31–38.

Green, M.F. (2007). Am. J. Psychiatry. *Cognition, drug treatment, and functional outcome in schizophrenia: a tale of two transitions* **164**, 992–994.

Hatcher, P.D., Brown, V.J., Tait, D.S., Bate, S., Overend, P., Hagan, J.J., and Jones, D.N. (2005). 5-HT_6 receptor antagonists improve performance in an attentional set shifting task in rats. *Psychopharmacology (Berl)* **181**, 253–259.

Idris, N., Neill, J., Grayson, B., Bang-Andersen, B., Witten, L.M., Brennum, L.T., and Arnt, J. (2010). Sertindole improves sub-chronic PCP-induced reversal learning and episodic memory deficits in rodents: involvement of 5-$HT_{(6)}$ and 5-HT (2A) receptor mechanisms. *Psychopharmacology (Berl)* **208**, 23–36.

Idris, N.F., Repeto, P., Neill, J.C., and Large, C.H. (2005). Investigation of the effects of lamotrigine and clozapine in improving reversal-learning impairments induced by acute phencyclidine and D-amphetamine in the rat. *Psychopharmacology (Berl)* **179**, 336–348.

Jentsch, J.D., and Roth, R.H. (1999). The neuropsychopharmacology of phencyclidine: from NMDA receptor hypofunction to the dopamine hypothesis of schizophrenia. *Neuropsychopharmacology* **20**, 201–225.

Keefe, R.S., Bilder, R.M., Davis, S.M., Harvey, P.D., Palmer, B.W., Gold, J.M., Meltzer, H.Y., Green, M.F., Capuano, G., Stroup, T.S., McEvoy, J.P., Swartz, M.S., Rosenheck, R.A., Perkins, D.O., Davis, C.E., Hsiao, J.K., and Lieberman, J.A. (2007). Neurocognitive effects of antipsychotic medications in patients with chronic schizophrenia in the CATIE trial. *Arch. Gen. Psychiatry* **64**, 633–647.

Kellendonk, C., Simpson, E.H., and Kandel, E.R. (2009). Modeling cognitive endophenotypes of schizophrenia in mice. *Trends Neurosci.* **32**, 347–358.

Kendall, I., Slotten, H. A., Codony, X., Burgueno, J., Pauwels, P. J., Vela, J. M., and Fone, K. C. (2010). E-6801, a 5-HT$_{(6)}$ receptor agonist, improves recognition memory by combined modulation of cholinergic and glutamatergic neurotransmission in the rat. *Psychopharmacology (Berl)* (Epub April 20). doi: 10.1007/s00213-010-1854-3.

King, M.V., Sleight, A.J., Woolley, M.L., Topham, I.A., Marsden, C.A., and Fone, K.C. (2004). 5-HT$_6$ receptor antagonists reverse delay-dependent deficits in novel object discrimination by enhancing consolidation – an effect sensitive to NMDA receptor antagonism. *Neuropharmacology* **47**, 195–204.

Kirby, B.P., Waddington, J.L., and O'Tuathaigh, C.M. (2009). Advancing a functional genomics for schizophrenia: psychopathological and cognitive phenotypes in mutants with gene disruption. *Brain Res. Bull.* **83**, 162–176.

Lacroix, L.P., Dawson, L.A., Hagan, J.J., and Heidbreder, C.A. (2004). 5-HT$_6$ receptor antagonist SB-271046 enhances extracellular levels of monoamines in the rat medial prefrontal cortex. *Synapse* **51**, 158–164.

Leng, A., Ouagazzal, A., Feldon, J., and Higgins, G.A. (2003). Effect of the 5-HT$_6$ receptor antagonists Ro04-6790 and Ro65-7199 on latent inhibition and prepulse inhibition in the rat: comparison to clozapine. *Pharmacol. Biochem. Behav.* **75**, 281–288.

Lewis, D.A., Hashimoto, T., and Morris, H.M. (2008). Cell and receptor type-specific alterations in markers of GABA neurotransmission in the prefrontal cortex of subjects with schizophrenia. *Neurotox. Res.* **14**, 237–248.

Lewis, D.A., and Moghaddam, B. (2006). Cognitive dysfunction in schizophrenia: convergence of gamma-aminobutyric acid and glutamate alterations. *Arch. Neurol.* **63**, 1372–1376.

Li, Z., Huang, M., Prus, A.J., Dai, J., and Meltzer, H.Y. (2007). 5-HT$_6$ receptor antagonist SB-399885 potentiates haloperidol and risperidone-induced dopamine efflux in the medial prefrontal cortex or hippocampus. *Brain Res.* **1134**, 70–78.

Lodge, D.J., Behrens, M.M., and Grace, A.A. (2009). A loss of parvalbumin-containing interneurons is associated with diminished oscillatory activity in an animal model of schizophrenia. *J. Neurosci.* **29**, 2344–2354.

Lodge, D.J., and Grace, A.A. (2009). Gestational methylazoxymethanol acetate administration: a developmental disruption model of schizophrenia. *Behav. Brain Res.* **204**, 306–312.

Loiseau, F., Dekeyne, A., and Millan, M.J. (2008). Pro-cognitive effects of 5-HT$_6$ receptor antagonists in the social recognition procedure in rats: implication of the frontal cortex. *Psychopharmacology (Berl)* **196**, 93–104.

McLean, S.L., Beck, J.P., Woolley, M.L., and Neill, J.C. (2008). A preliminary investigation into the effects of antipsychotics on sub-chronic phencyclidine-induced deficits in attentional set-shifting in female rats. *Behav. Brain Res.* **189**, 152–158.

Minabe, Y., Shirayama, Y., Hashimoto, K., Routledge, C., Hagan, J.J., and Ashby Jr., C.R. (2004). Effect of the acute and chronic administration of the selective 5-HT$_6$ receptor antagonist SB-271046 on the activity of midbrain dopamine neurons in rats: an in vivo electrophysiological study. *Synapse* **52**, 20–28.

Mitchell, E.S., and Neumaier, J.F. (2005). 5-HT$_6$ receptors: a novel target for cognitive enhancement. *Pharmacol. Ther.* **108**, 320–333.

Mitchell, E.S., and Neumaier, J.F. (2008). 5-HT$_6$ receptor antagonist reversal of emotional learning and prepulse inhibition deficits induced by apomorphine or scopolamine. *Pharmacol. Biochem. Behav.* **88**, 291–298.

Mohn, A.R., Gainetdinov, R.R., Caron, M.G., and Koller, B.H. (1999). Mice with reduced NMDA receptor expression display behaviors related to schizophrenia. *Cell* **98**, 427–436.

Mork, A., Karlsson, J. J., and Arnt, J. (2010). The 5-HT_6 receptor antagonist, Lu AE58054 blocks the effect of 5-HT_6 receptor-stimulated modulation of extracellular neurotransmitter levels in the medial prefrontal cortex of rats. Society of Neuroscience Abstracts, 405.28.

Mork, A., Witten, L.M., and Arnt, J. (2009). Effect of sertindole on extracellular dopamine, acetylcholine, and glutamate in the medial prefrontal cortex of conscious rats: a comparison with risperidone and exploration of mechanisms involved. *Psychopharmacology (Berl)* **206**, 39–49.

Moser, P.C., Hitchcock, J.M., Lister, S., and Moran, P.M. (2000). The pharmacology of latent inhibition as an animal model of schizophrenia. *Brain Res. Brain Res. Rev.* **33**, 275–307.

Olsen, C.K., Brennum, L.T., and Kreilgaard, M. (2008). Using pharmacokinetic-pharmacodynamic modelling as a tool for prediction of therapeutic effective plasma levels of antipsychotics. *Eur. J. Pharmacol.* **584**, 318–327.

Penschuck, S., Flagstad, P., Didriksen, M., Leist, M., and Michael-Titus, A.T. (2006). Decrease in parvalbumin-expressing neurons in the hippocampus and increased phencyclidine-induced locomotor activity in the rat methylazoxymethanol (MAM) model of schizophrenia. *Eur. J. Neurosci.* **23**, 279–284.

Pitsikas, N., Zisopoulou, S., Pappas, I., and Sakellaridis, N. (2008). The selective 5-$HT_{(6)}$ receptor antagonist Ro 04-6790 attenuates psychotomimetic effects of the NMDA receptor antagonist MK-801. *Behav. Brain Res.* **188**, 304–309.

Pouzet, B., Didriksen, M., and Arnt, J. (2002). Effects of the 5-$HT_{(6)}$ receptor antagonist, SB-271046, in animal models for schizophrenia. *Pharmacol. Biochem. Behav.* **71**, 635–643.

Reynolds, G.P., Abdul-Monim, Z., Neill, J.C., and Zhang, Z.J. (2004). Calcium binding protein markers of GABA deficits in schizophrenia – postmortem studies and animal models. *Neurotox. Res.* **6**, 57–61.

Roberts, J.C., Reavill, C., East, S.Z., Harrison, P.J., Patel, S., Routledge, C., and Leslie, R.A. (2002). The distribution of 5-$HT_{(6)}$ receptors in rat brain: an autoradiographic binding study using the radiolabelled 5-$HT_{(6)}$ receptor antagonist [(^{125})I]SB-258585. *Brain Res.* **934**, 49–57.

Rodefer, J.S., Murphy, E.R., and Baxter, M.G. (2005). PDE10A inhibition reverses subchronic PCP-induced deficits in attentional set-shifting in rats. *Eur. J. Neurosci.* **21**, 1070–1076.

Rodefer, J.S., Nguyen, T.N., Karlsson, J.J., and Arnt, J. (2008). Reversal of subchronic PCP-induced deficits in attentional set shifting in rats by sertindole and a 5-HT_6 receptor antagonist: comparison among antipsychotics. *Neuropsychopharmacology* **33**, 2657–2666.

Sams-Dodd, F. (1999). Phencyclidine in the social interaction test: an animal model of schizophrenia with face and predictive validity. *Rev. Neurosci.* **10**, 59–90.

Schechter, L.E., Lin, Q., Smith, D.L., Zhang, G., Shan, Q., Platt, B., Brandt, M.R., Dawson, L.A., Cole, D., Bernotas, R., Robichaud, A., Rosenzweig-Lipson, S., and Beyer, C.E. (2008). Neuropharmacological profile of novel and selective 5-HT_6 receptor agonists: WAY-181187 and WAY-208466. *Neuropsychopharmacology* **33**, 1323–1335.

Schotte, A., Janssen, P.F., Gommeren, W., Luyten, W.H., Van Gompel, P., Lesage, A.S., De Loore, K., and Leysen, J.E. (1996). Risperidone compared with new and reference antipsychotic drugs: in vitro and in vivo receptor binding. *Psychopharmacology (Berl)* **124**, 57–73.

Schreiber, R., Vivian, J., Hedley, L., Szczepanski, K., Secchi, R.L., Zuzow, M., van Laarhoven, S., Moreau, J.L., Martin, J.R., Sik, A., and Blokland, A. (2007). Effects of the novel 5-$HT_{(6)}$ receptor antagonist RO4368554 in rat models for cognition and sensorimotor gating. *Eur. Neuropsychopharmacol.* **17**, 277–288.

Skarsfeldt, T. (1992). Electrophysiological profile of the new atypical neuroleptic, sertindole, on midbrain dopamine neurones in rats: acute and repeated treatment. *Synapse* **10**, 25–33.

Skarsfeldt, T. (1996). Differential effect of antipsychotics on place navigation of rats in the Morris water maze. A comparative study between novel and reference antipsychotics. *Psychopharmacology (Berl)* **124**, 126–133.

Snigdha, S., and Neill, J.C. (2008a). Efficacy of antipsychotics to reverse phencyclidine-induced social interaction deficits in female rats – a preliminary investigation. *Behav. Brain Res.* **187**, 489–494.

Snigdha, S., and Neill, J.C. (2008b). Improvement of phencyclidine-induced social behaviour deficits in rats: involvement of 5-HT_{1A} receptors. *Behav. Brain Res.* **191**, 26–31.

Tseng, K.Y., Chambers, R.A., and Lipska, B.K. (2009). The neonatal ventral hippocampal lesion as a heuristic neurodevelopmental model of schizophrenia. *Behav. Brain Res.* **204**, 295–305.

Upton, N., Chuang, T.T., Hunter, A.J., and Virley, D.J. (2008). 5-HT_6 receptor antagonists as novel cognitive enhancing agents for Alzheimer's disease. *Neurotherapeutics* **5**, 458–469.

Wang, C.Z., Yang, S.F., Xia, Y., and Johnson, K.M. (2008). Postnatal phencyclidine administration selectively reduces adult cortical parvalbumin-containing interneurons. *Neuropsychopharmacology* **33**, 2442–2455.

Ward, R.P., and Dorsa, D.M. (1996). Colocalization of serotonin receptor subtypes 5-HT_{2A}, 5-HT_{2C}, and 5-HT_6 with neuropeptides in rat striatum. *J. Comp. Neurol.* **370**, 405–414.

Woolley, M.L., Marsden, C.A., and Fone, K.C. (2004). 5-HT_6 receptors. *Curr. Drug Targets* **3**, 59–89.

Wünsch, A. M., Nguyen, H. Q., Guynn, L., and Rodefer, J. S. (2006). Effects of 5-HT_6 antagonists on attenuating PCP-induced cognitive deficits in a rodent set-shifting (ED/ID) task. Society of Neuroscience Abstracts 264.23.

Zhang, W., and Bymaster, F.P. (1999). The in vivo effects of olanzapine and other antipsychotic agents on receptor occupancy and antagonism of dopamine D_1, D_2, D_3, $5HT_{2A}$ and muscarinic receptors. *Psychopharmacology (Berl)* **141**, 267–278.

5-HT_6 RECEPTOR LIGANDS AS ANTIDEMENTIA DRUGS

Ellen Siobhan Mitchell

Unilever Research and Development, Vlaardingen 3135 XB, The Netherlands

I. Introduction

Age-related dementia is a huge public health issue that will undoubtedly increase along with the average age of the world's population. Because of the

INTERNATIONAL REVIEW OF
NEUROBIOLOGY, VOL. 96
DOI: 10.1016/B978-0-12-385902-0.00007-3

huge financial and emotional burdens dementia causes, pharmaceutical, biotechnology, and academic labs are continually increasing their efforts toward effective therapies using a wide range of targets. Three main approaches have been conceived for antidementia drugs: those that are curative, which actually reverse pathology and brain dysfunction causing dementia, those that are palliative, which delay progression of disease but do not cure it, and preventative treatments, which avert pathological changes but are ineffective when pathology has already occurred.

Despite the huge impetus for breakthroughs, only a few drugs are currently approved for Alzheimer's patients, and almost all of them have the same mechanism of acetylcholinesterase inhibition, which may only be palliative and not curative. Pharmaceutical companies have so far been wholly unsuccessful in bringing to market a disease-modifying therapy for many reasons: mainly because of the many heterogeneous causes of neurodegeneration, but also because of the many unforeseen, toxic side effects of drugs that reverse or block putative pathological pathways such as amyloid processing. Pharmaceutical researchers are currently revisiting palliative therapies as less risky avenues of drug development. 5-HT_6 ligands are attractive propositions for treating dementia for several reasons: (1) almost all 5-HT_6 antagonists have been shown to improve memory performance in aged or cholinergically compromised animals (Mitchell and Neumaier, 2005); (2) 5-HT_6 antagonists appear to have little toxicity or negative side effects on physiology or brain function (Upton *et al.*, 2008); (3) 5-HT_6 receptors are found almost wholly in the brain, and thus "most antagonists" are inactive on peripheral targets, decreasing chance of side effects (Geldenhuys and Van der Schyf, 2009). (4) Since 5-HT_6 antagonists improve memory via a different mechanism than current dementia treatments, they can be used as adjunct treatments. A final emerging, although not yet validated, reason for developing 5-HT_6 drugs is the possibility of curative as well as palliative actions, and this will be addressed later in this chapter.

Despite the above-mentioned positive attributes relatively few 5-HT_6 antagonists have progressed to clinical trials. This chapter will review the most current research on 5-HT_6 mechanisms for attenuating dementia, the status of drugs in clinical development, and controversial issues of 5-HT_6 drug efficacy, most notably the contradictory procognitive data for both 5-HT_6 antagonists and agonists.

5-HT_6 receptors are found in many brain regions and affect many neurotransmitter systems including acetylcholine, norepinephrine, dopamine, glutamate, and GABA (Mitchell and Neumaier, 2005). Because they are most abundantly expressed in GABAergic neurons of the striatum, nucleus accumbens, cortex, and hippocampus (Roberts *et al.*, 2002), 5-HT_6 receptors likely modulate cognition via GABAergic signaling in basal ganglia and limbic circuits.

II. Studies Using Animal Models of Dementia

A. Pharmacological Models (Anticholinergic and Antiglutamatergic)

One of the hallmarks of Alzheimer's disease, along with amyloid plaque and tau tangle formation, is the attenuation of acetylcholine signaling (Sambeth *et al.*, 2007). Thus, cholinergic antagonists are used to model this deficiency in order to evaluate putative antidementia drugs.

The most predominantly used model of disrupted cholinergic function is scopolamine administration before or after a learning task. Scopolamine, a M1 muscarinic antagonist, causes amnesia in young animals at moderate doses (0.2–0.4 mg/kg) and in older animals at smaller doses (0.1–0.2 mg/kg). Dose must be carefully monitored, since scopolamine can cause sedation and decreases in motor activity at higher concentrations.

A yawning and stretching phenomenon caused by the 5-HT_6 antagonist, RO 046790, implied a cholinergic influence of the 5-HT_6 receptors (Bourson *et al.*, 1995). However, the first pharmacological demonstration of cholinergic-mediated procognitive effects of 5-HT_6 antagonists was reported (Meneses, 2001a), using autoshaping, an instrumental learning task. RO 046790 (10 mg/kg i.p.) improved learning consolidation alone and also ameliorated poor performance due to scopolamine (0.17 mg/kg i.p.). Following this report, Woolley *et al.* (2003) demonstrated that RO 046790 reversed the amnesic effects of scopolamine (but not methyl-scopolamine) in a novel object recognition task, but had no effect on poor memory consolidation due to administration of the D2 dopamine antagonist, raclopride. Another 5-HT_6 antagonist, SB-271046, reversed scopolamine deficit in fear conditioning tasks such as the passive avoidance task (Foley *et al.*, 2004; Leng *et al.*, 2003). More recently another 5-HT_6 antagonist, SB-742457, improved performance on a delayed nonmatch to sample (DNMS) task, which tests spatial working memory (Callaghan *et al.*, 2009).

In fact, many different types of cognition compromised by scopolamine have been shown to be rescued by 5-HT_6 antagonists including latent inhibition, fear potentiated startle, social recognition, and Morris water maze (Costa-Aze *et al.*, 2009; Foley *et al.*, 2004; Mitchell and Neumaier, 2008; Mitchell *et al.*, 2006). Thus, it can be concluded that 5-HT_6 blockade is highly effective in cholinergic models of dementia; however, since AD is characterized by loss of cholinergic afferents, lesions of the nucleus basalis of Meynert may be more appropriate for modeling pathology and assessing 5-HT_6 ligands' translational potential.

Interestingly, many studies using 5-HT_6 antagonists in uncompromised animals (i.e., without scopolamine-induced amnesia) did not report improvements in visual, episodic memory tasks such as novel object recognition (Hirst *et al.*, 2006; King *et al.*, 2004, 2006b), while all but one study reported enhanced autoshaping performance with 5-HT_6 antagonists (Lindner *et al.*, 2003; Liy-Salmeron and Meneses, 2007; Meneses, 2001a, 2001b; Nelson, 2005). Such results suggest a stronger influence of 5-HT_6 blockade on tasks involving the striatum, where 5-HT_6 receptor expression is high, versus tasks primarily involving the hippocampus, where 5-HT_6 receptor expression is moderate (Roberts *et al.*, 2002). Nevertheless, the above-mentioned results demonstrate the wide range of 5-HT_6 antagonists on diverse cognitive tasks and functions.

Another class of drugs used to induce cognitive deficits acts via glutamatergic blockade. MK-801, an NMDA antagonist, is a commonly used antiglutamatergic drug, as is ketamine, another NMDA antagonist with less hypnotic effects (Kelland *et al.*, 1993). Serotonin-6 ligands have been less successful in alleviating antiglutamatergic models of dementia (King *et al.*, 2004; Pitsikas *et al.*, 2008). Several studies have reported significant rescue of MK-801-induced learning impairment in novel object recognition and Morris water maze (King *et al.*, 2004; Marcos *et al.*, 2008; Pitsikas *et al.*, 2008). However, motor activity disturbances have sometimes been associated with MK-801 administration (Ahlander *et al.*, 1999), and may be responsible for the learning impairment rather than direct changes on memory function. Ketamine (0.1 mg/kg i.p.) was used to induce memory deficits in an autoshaping task, and these deficits were reversed by administration of SB-399885 directly into the hippocampus, cortex, or systemically (10 mg/kg i.p.) (Liy-Salmeron and Meneses, 2007, 2008). These studies demonstrate that 5-HT_6 antagonists may be ineffective against sedative-induced learning deficits but do reverse the amnesic effects of direct NMDA inhibition in the cortex and hippocampus.

Serotonin-6 ligands have been assessed with several other types of pharmacologically induced memory impairment, such as chemical lesioning of 5-HT afferents or drugs that decrease serotonin or dopamine signaling. However, since these methods are not traditionally used for modeling dementia they are beyond the scope of this chapter.

B. Aging Animal Models

Animal models using chemically induced amnesia do not capture the time-delayed and environmental onset of human neurodegeneration. Moreover, since most amnesic drugs have highly defined thresholds of effectiveness it is difficult to evaluate premorbid or presymptomatic aspects of dementia. Thus, it is also expedient to use aged animals in assessing procognitive effects of 5-HT_6 ligands.

To date, only one study has thoroughly examined 5-HT_6 blockade on rodent learning performance at several different life stages. Costa *et al.* (2009) used young (3 months), mature (12 months), middle-aged (18 months), and old (21 months) mice to evaluate the procognition effects of the 5-HT_6 antagonist, SB-271046,

(10 mg/kg i.p.) in a Y maze. During the first trial, an animal is placed in a starting arm and allowed to explore, although one of the two other arms is closed. During the second trial, when all arms are accessible, an animal's place memory is assessed by measuring the amount of time spent in the previously blocked arm. Using progressively longer intertrial intervals of 30, 60, 120, 240, and 360 min, saline-treated mice retained memory of the novel arm up to 120 min later. However, young mice given SB-271046 explored the novel arm more at 240 and 360 min. Aged mice (21 months) had significant recall of the maze 60 min later and SB-240146 improved acquisition memory in an ITI of 240 min, but not 360 min.

The Y maze task allows several types of memory to be measured: acquisition, consolidation, and retrieval. Like King *et al.* (2004) reported previously, Costa *et al.* (2009) found that SB-271046 administered postconsolidation was not as effective as administration immediately before or after learning, and moreover this observation held for several age groups.

Chronic 5-HT$_6$ antagonist treatment has also produced encouraging results in aged animals SB-399885 (10 mg/kg). Foley *et al.* (2004) treated 18-month-old rats with SB-271046 (10 mg/kg i.p.) for 40 days and found that their ability to navigate a Morris water maze was better than age-matched controls. Similarly Mitchell *et al.* (2009), examined the effects of 7-day treatment RO 4368554 (5 mg/kg i.p.) on 15- to 18-month rats' performance on a social recognition task, where rats showed better recall of novel juveniles than controls. Interestingly, when Mitchell and colleagues gave daily injections of RO 4368854 for 2 weeks and then tested the rats' memory 1 day after treatment had ended, the rats still performed significantly better than vehicle-treated rats. This finding indicates that long-term treatment causes changes in brain function that allow for better learning even without the drug on board. This topic will be explored further in the biomarkers section below.

Serotonin-6 antagonists appear to improve working memory, as well as long-term memory, in aged animals and indeed working memory may be more sensitive to 5-HT$_6$ inhibition. Recently the operant DNMS test, a working memory task, and a modified version of the Morris water maze were used to test middle-aged and young rats given SB-742457 (Callaghan *et al.*, 2009). Following a 6-week training protocol in the DNMS task, both young (3 months) and middle-aged (12–14 months) rats received chronic treatment (3 mg/day) of SB-742457, with the results that the drug-treated middle-aged group matched the performance of the young control group. In the maze task animals received treatment (3 mg/day) of SB-742457 24 h before testing, but no change was seen in performance as compared to controls.

C. Summary of Animal Models

5-HT$_6$ antagonists are effective against learning impairment from anticholinergic and antiglutamatergic models of dementia. Due to the purported low expression of 5-HT$_6$ receptors in the murine brain (Hirst *et al.*, 2003), few studies have employed this species. Earlier studies using mice found no procognition effects of Lindner *et al.* (2003), although more recent studies have demonstrated that new 5-HT$_6$ ligands do appear to be therapeutically active (Costa *et al.*, 2009; Schechter *et al.*, 2008). Indeed, later expression studies also showed abundant 5-HT$_6$ receptor expression in mice (Bibancos *et al.*, 2007). It is not clear how to resolve the contradictory findings, which may be due to various mouse strains showing differential 5-HT$_6$ receptor expression. In any case, it is clear that researchers must tread very carefully in evaluating preclinical efficacy of 5-HT$_6$ antagonists in mice.

Since the most relevant Alzheimer's disease models are transgenic mice, it is surprising that no reports are available on 5-HT$_6$ ligands in, for example, an amyloid precursor protein (APP) overexpressing mouse. However, APP overexpression and other gene mutations in mice do not model sporadic dementia, and are more useful for assessing therapies that target the amyloid system. In any case, the lack of transgenic mouse data has not hindered the progress of 5-HT$_6$ ligands.

There are other issues with predicting how effective 5-HT$_6$ ligands will be in humans based on current animal data. For instance, rodents sleep during light hours and are active during dark periods, and yet drugs are often tested during their inactive times. Disturbances in rats' natural circadian rhythm may cause inconsistent performance. Moreover, there is also evidence that the 5-HT$_6$ antagonist RO 4368554 increases sleep duration and decreases waking in rats when given during the middle of the dark (active) period (Morairty *et al.*, 2008). Since Alzheimer's disease is associated with sleep problems, which may be affected by 5-HT$_6$ antagonists, it is important to understand the possible circadian effects from 5-HT$_6$ blockade. Another interesting issue is the weight-decreasing effects of some 5-HT$_6$ ligands (Halford *et al.*, 2007; Heal e*t al.*, 2008), which may present problems in some demented patients, who often have poor appetite.

Dementia progression in humans is a highly variable process that demands many types of animal models to assess the diverse pathological mechanisms at work. As seen in the section above, only a select number of animal models of dementia have been employed to study the cognitive effects of 5-HT$_6$ ligands. Because of this limitation, it is difficult to predict the efficacy of 5-HT$_6$ ligands on, for example, vascular dementia, which requires animal models exhibiting infarcts and oxidative damage.

III. Biomarkers of 5-HT$_6$ Receptors' Antidementia Effects

This section gives an overview of signaling molecules and receptors that have been implicated in 5-HT$_6$ mechanism of action. Since many of these molecules are also associated with learning and memory, it may be possible to predict clinical efficacy with these putative "biomarkers."

A. Fyn

Fyn, a Src family tyrosine kinase with roles in aging pathogenesis, forms complexes with 5-HT_6 receptors, which mediates phosphorylation of extracellular signal-related kinase-1/2 (ERK1/2) (Yun *et al.*, 2007). Part of the phosphokinase A (PKA) signaling pathway, Fyn has been implicated in key molecular changes during memory consolidation, most notably via regulation of synaptic growth (Braithwaite *et al.*, 2006; Lim *et al.*, 2009). Interestingly, overexpression of APP causes synaptotoxicity via Fyn (Chin *et al.*, 2005). Conversely, Fyn also has been implicated in establishing fear conditioned learning, since Fyn-knockout mice fail to acquire fear conditioning, while in wild-type mice Fyn is upregulated immediately after contextual fear (Isosaka *et al.*, 2008).

Because of Fyn's exacerbation of neurodegenerative mechanisms, inhibition of this pathway via chronic 5-HT_6 antagonist treatment may provide therapeutic, not just palliative, benefits. No studies have yet reported on whether 5-HT_6 antagonists affect Fyn activity during normal or pathological conditions. However, it must also be noted that Fyn recruits NMDA receptors to the synapse and has roles in long-term potentiation (LTP) (Pena *et al.*, 2010; Xu *et al.*, 2006).

It still remains to be seen what the functional implication of Fyn-5-HT_6 complexes may be in terms of memory formation. Fyn has been associated to worsening cognitive pathology in Alzheimer's patients' brains, and inhibition of Fyn is currently being investigated as a treatment for dementia (Braithwaite *et al.*, 2006). However, it is possible that Fyn's direct association with 5-HT_6 receptors may cause it to be less active in mediation of neurotoxic pathways; further studies will likely reveal whether this is the case.

B. BDNF and Arc

The brain-derived growth factor (BDNF) is one of the most extensively studied molecules in neuronal growth and memory formation, and is also a target for neurodegenerative disease therapy. BDNF protein expression as well as Arc, an immediate early gene, has been reported to be increased after acute administration of a 5-HT_6 agonist, LY-586713 (1 mg/kg i.p.), in the hippocampus (De Foubert *et al.*, 2007). This effect is specific to 5-HT_6 activity, since 5-HT_6 antagonist administration blocked BDNF expression. However, LY-586713 at a higher dose (10 mg/kg) had no effect on BDNF expression. It is not immediately apparent how 5-HT_6 receptor activation upregulates BDNF and how this reconciles with the fact that 5-HT$_6$ blockade is procognitive while 5-HT$_6$ stimulation is generally not. Nonetheless, patents have been filed for 5-HT$_6$ agonists as neurodegenerative therapeutics specifically for their ability to induce BDNF. A follow-up study on what types of neurons exhibit increased BDNF expression after 5-HT$_6$ stimulation may shed more light on therapeutic applicability.

C. Jun-Activating Domain Binding Protein 1 (Jab1)

The Jun-activating domain binding protein 1 (Jab1) has been studied extensively for its interactions with hypoxia-inducible factor (HIF)-1alpha. Recently, a paper suggested that Jab1 has a role in preventing neurodegeneration, since suppression of Jab1 decreases HIF-1alpha activity during inflammation, causing downturns in VEGF (Lungu *et al.*, 2008). Interestingly, downregulation of Jab1 expression decreases 5-HT$_6$ receptor expression, an effect also observed after memory training in rats (Yun *et al.*, 2010). Conversely, stimulation of 5-HT$_6$ receptors induces the translocation of Jab1 into the cell nucleus and increases c-Jun phosphorylation. Lastly, in this same paper, 5-HT$_6$ receptors and Jab1 were also shown to be upregulated in middle cerebral artery occlusion (MCAO)-induced focal cerebral ischemic rats and in hypoxic cultured cells, suggesting possible protective roles for 5-HT$_6$ receptors and Jab1.

D. 5-HT$_6$ Receptor Expression

Considering that inhibition of 5-HT$_6$ receptors generally produce procognitive effects especially in aged animals, what is the functional relevance of variations in endogenous receptor levels? Given that blockade of 5-HT$_6$ receptors enhances memory; would decreased receptor expression signify better learning capacity? To date two studies have examined the functional role of 5-HT$_6$ receptor expression in human brains. Garcia-Alloza *et al.* (2004) correlated serotonin receptor expression patterns in the brains of Alzheimer's patients to cognitive and noncognitive symptoms reported during treatment. As measured by the Present Behavioral Examination (PBE), hyperactive behavior was associated to lower temporal cortex 5-HT$_6$ receptor density, and aggressive behavior was associated to higher ratios of 5-HT$_6$ receptors to cholinacetyltransferase (ChAT) in the frontal cortex. No correlation was found for mental ability as measured by the Mini Mental State Examination (MMSE), which may be due to the low sensitivity of the MMSE in detecting subtle changes in cognition. A more recent study examined 5-HT$_6$ receptor expression and activity in postmortem Alzheimer's patients' brains. Cortical 5-HT$_6$ receptor density was lower in AD brains as compared to controls, and cAMP accumulation after direct stimulation with a 5-HT_6 agonist, E-6801, was also decreased (Marcos *et al.*, 2009). Reduced cAMP accumulation was correlated with increased psychotic symptoms as reported by patient caretakers, most notably in females. These above-mentioned results indicate more promising applications for schizophrenia treatment than dementia. However, many Alzheimer's patients also exhibit behavioral and psychological syndromes of dementia (BPSD) such as depression, agitation, and poor sleep. Thus, it may be concluded that 5-HT_6 antagonists have potential for treating BPSD.

Several studies have examined changes in 5-HT_6 receptor expression in animals during aging, learning, and drug administration. Successive days of Morris water maze training decreased 5-HT_6 receptor protein and mRNA receptor expression in the rat hippocampus (Marcos *et al.*, 2009). The functional implication of this is not clear, but it does suggest a temporary feedback mechanism in which to enhance learning by receptor downregulation. However, this finding reflects changes in expression after just one type of learning in just one brain region, and did not investigate whether expression eventually goes back to baseline. This same study reported that systemic SB-271046 increased phosphorylated cAMP response element binding protein-1 (CREB-1), a signaling factor strongly linked to memory consolidation and plasticity, but not CREB-2 (Marcos *et al.*, 2009). While pCREB-1 did not increase after training, there was an additive effect of learning on SB-271046-induced pERK1/2. A more recent study examined the effects of striatum-based learning and acute 5-HT_6 antagonist treatment; autoshaping training and SB-399885 increased 5-HT_6 receptor mRNA expression in the striatum and cortex (Huerta-Rivas *et al.*, 2010). Thus, the duration of 5-HT_6 blockade and the type of learning task seem to regulate 5-HT_6 receptors differentially, although it must be noted that mRNA expression does not always completely correspond to protein expression.

E. PSA-NCAM

The polysialated neural cell adhesion molecule (PSA-NCAM) is involved in structural plasticity initiated by LTP. Chronic 5-HT_6 antagonist treatment (SB-271046) increases PSA-NCAM in the cortex and subgranular dentate gyrus (Foley *et al.*, 2008). Upton *et al.* (2008) recently suggested that upregulation of PSA-NCAM via 5-HT_6 blockade may be occurring via inhibition of GABA release on glutamatergic neurons in the hippocampus. A subsequent increase in glutamatergic signaling would in turn promote the survival of neural progenitors, whose expression of PSA-NCAM increases as they mature into functioning neurons. Since PSA-NCAM is necessary for synaptogenesis and synaptic remodeling, such young neurons may be the primary mediators of strong memory traces during learning. If such a proposed mechanism can be more conclusively demonstrated (i.e., with quantification and colocalizaton of immature neuronal markers and PSA-NCAM in treated animals) then a stronger case is built for long-term functional changes with 5-HT$_6$ antagonist therapy. Indeed, as reported in a conference abstract, another 5-HT$_6$ antagonist, RO 4899237-002, increased LTP in the CA1 region but not in the dentate gyrus, and this same concentration was also shown to decrease LTD in the CA1 region (Robillard and Christie, 2007). However, the finding that 5-HT$_6$ antagonist treatment may decrease 5-HT$_6$ receptor expression in the hippocampus (Huerta-Rivas *et al.*, 2010) suggests that PSA-NCAM expression may not be sustained over time.

F. mTOR

The mammalian target of rapamycin (mTOR) has recently garnered ample attention as a regulator of protein synthesis at the translational level. Interestingly, the inhibition of mTOR is a potential target mechanism for antiaging therapies, via more efficient processing of proteins, thereby preventing deposition of protein fragments such as beta-amyloid. In an abstract, Meffre *et al.* (2010) reported that the 5-HT$_6$ receptor agonist, WAY-181187, increased phosphorylation of mTOR (Ser2448) in the striatum and prefrontal cortex. Two mTOR-related signaling factors, phosphatidylinositol 3-kinase catalytic subunit type 3 and neurofibromin, were also reported to be activated by 5-HT$_6$ stimulation. WAY-181187 also increased phosphorylation of a downstream target of mTOR, the ribosomal protein S6, which was abolished by the 5-HT$_6$ receptor antagonist, SB-258585. Rapamycin, the well-characterized inhibitor for which mTOR is named, prevented the decrements in social recognition memory elicited by WAY-181187 administration, giving further credence to the amnesic effects of 5-HT$_6$ receptor activation via mTOR. It should be noted that studies have reported controversial findings of rapamycin inhibiting LTP and memory consolidation (Blundell *et al.*, 2008), while other studies found no effect of rapamycin on memory (Hoeffer and Klann, 2010). Nonetheless, careful gauging of mTOR inactivation via 5-HT$_6$ agonists may be required to elicit antiaging benefits without detriment to cognition.

An intriguing implication from the above-mentioned study is the possible use of 5-HT$_6$ antagonists for inhibition of mTOR, which suggests that therapeutic effects may not simply be temporary changes in acetylcholine or glutamate signaling, but also long-term protection against aging neurodegeneration through inefficient protein synthesis. Rapamycin itself is not an appropriate preventative agent, due to side effects such as those listed above. However, 5-HT$_6$ antagonists may offer a brain-specific intervention for more efficient regulation of protein synthesis, and preserved neuronal function.

G. Summary

The molecules discussed above are emerging biomarkers of clinical efficacy in translational models of dementia such as aged rodents. The best marker of clinical utility is improved cognition, yet it is increasingly important for drug developers to also pinpoint mechanistic indicators of efficacy. Additionally, since many behavioral measures may give inconsistent results due to poorly chosen parameters, high signal-to-noise ratio, or inappropriate baseline, biomarkers can also provide clues on what types of cognition tests should be used and also what type of subjects may benefit best from treatment. Due to the difficulty of measuring brain-based signaling factors, it is unlikely that any of the above-mentioned molecules would be used to assess efficacy in humans. However, this section does highlight potential biomarkers for future 5-HT_6 ligand testing.

IV. Clinical Studies of 5-HT_6 Antagonists

The number of 5-HT_6 antagonists in clinical studies is quite large, considering the fact that most drugs currently being developed are curative, and 5-HT_6 antagonists' proposed action is palliative. However, only one Phase II clinical trial of a 5-HT_6 antagonist has been reported in detail (Maher-Edwards *et al.*, 2009), and the results show SB-742457 to be as efficacious as donepezil, an acetylcholinesterase inhibitor that is currently the most prescribed drug for Alzheimer's dementia. However, the clinical significance of these results was mixed and it is still not clear what potential 5-HT_6 ligands will have for Alzheimer's patients. Still, 5-HT_6 antagonists may have the potential to augment donepezil and several clinical trials are currently assessing 5-HT_6 antagonists as adjunct therapy for Alzheimer's patients currently taking acetylcholinesterase inhibitors, and these results are eagerly awaited. This section describes 5-HT_6 antagonists currently in clinical studies.

A. Avineuro

AVN-322 is one of the leading molecules for Avineuro, a biotechnology company in California. In animal studies AVN-322 was shown to be as effective as memantine, tacrine, SB-742457, and PRX-07034 in restoring performance after scopolamine or MK-801 administration (Tkachenko *et al.*, 2009). AVN-322 significantly restores both scopolamine- and MK-801-induced cognitive dysfunction. In a Phase I trial Avineuro's 5-HT_6 antagonist, AVN-322, demonstrated good tolerability and no adverse event in a range of dose levels (ClinicalTrials.gov, 2010). Plans for a Phase II trial are underway and it is expected to begin in late 2010.

B. Biovitrum/Proximagen

Biovitrum has run several Phase I studies on BVT.5182, which was found to be safe and well tolerated (Garfield and Heisler, 2009). While Biovitrum is currently focused on antiobesity uses of 5-HT_6 ligands, it recently sold clinical development rights for cognition applications to Proximagen (http://www.proximagen.com/RNS30-10-09.shtml). One of Biovitrum's 5-HT_6 antagonists will likely be chosen for Phase II trials for antidementia efficacy sometime in the next 3 years.

C. Epix

Epix's leading 5-HT_6 antagonist, PRX-07034, is a novel arylpiperazine with potent antagonism (K_i = 4.0 nM) and good preclinical efficacy (Upton *et al.*, 2008). Specifically, PRX-07034 improved memory consolidation in novel object recognition in rats by itself and in conjunction with donepezil (King *et al.*, 2006a). PRX-07034 also improved cognitive functioning in both aged and scopolamine-treated animals in tasks assessing working memory and executive function, as well as long-term memory. However, PRX-07034 (10 mg/kg i.p.) failed to reverse impaired instrumental learning in rats overexpressing striatal 5-HT_6 receptors, which was reversed by SB-258585 (5 mg/kg i.p.). Moreover, PRX-07034 did not enhance instrumental learning in normal rats (unpublished data, Neumaier lab). It is difficult to understand why PRX-07034 failed to have a specific effect on a 5-HT_6-mediated change in instrumental learning when a purportedly less bioavailable ligand was effective.

Only Phase I studies had been completed so far for PRX-07034, where it proved to be safe and moderately appetite-suppressing. A 28-day Phase Ib clinical trial with 33 moderately obese, but not cognitively impaired, adults showed significant improvement with a Cogscreen measurement tool.

Cogscreen is a computer-administered and scored testing battery of attention, memory, reaction time, and problem solving, using the following subtests: Backward Digit Span (BDS) traditional math word problems, Visual Sequence Comparison (VSC), Symbol Digit Coding (SDC), Matching to Sample (MTS) recognition of patterns, and Manikin, which measures mental rotation. Subjects are also tested with a Divided Attention Task (DAT) and Auditory Sequence Comparison (ASC), a visual sequencing and scanning task that requires respondents to sequence numbers, letters, and an alternating set of numbers and letters.

Subjects' speed and accuracy, as measured by a Cogscreen pooled endpoint parameter, was improved significantly with a 600 mg ×2 daily dose of PRX-07034, and thus this finding is the only indication that 5-HT_6 antagonists may act as cognition enhancers in healthy people. Interestingly, Cogscreen AE, an aeronautical pilot performance version, has been used to demonstrate efficacy of so-called cognition enhancers such as Adderall™ (Kay, 2004), which purportedly improve performance in mentally and physically healthy adults (de Jongh *et al.*, 2008). In the Epix Phase I trial, Cogscreen was used as a secondary endpoint in clinically healthy, yet obese adults. However, obesity is increasingly linked to impaired cognition. Thus, it remains to be seen whether improvement seen was due to weight loss or specific 5-HT_6 effects.

In 2009, Epix Pharmaceuticals closed its labs due to lack of funding. Its catalog of drugs in clinical trials was sold at auction. The development status of PRX-07034 is currently unknown.

D. GlaxoSmithKline SB-742457

GlaxoSmithKline has finished two Phase II trials to study the effects of SB-742457 and has completed recruitment for two more.

The main finding of the recently completed Phase II study was that SB-742457 performed as well as donepezil in key cognition parameters. This is important since, if approved, SB-742457 will likely compete for market share with donepezil (marketed as Aricept by Pfizer). Thus, SB-742457 was tested against donepezil, as well as in a monotherapy study using endpoints from the original clinical trials for donepezil.

Specifically, SB-74257 performed as well as donepezil in the Clinician's Interview-Based Impression of Change-plus Caregiver Input (CIBIC+) after 24-week treatment. SB-742457 alone (15, 35 mg) or added to donepezil treatment produced moderate results on several parameters of the ADAS-Cog in subjects with mild-to-moderate AD (MMSE 12–24). A 5 mg dose had no significant effects. Overall cognitive improvement from SB-742457 treatment was statistically significant, but clinically insignificant. Additionally, there was no separation between drug and placebo treatments on the specific ADAS-Cog scale. However, a subgroup with moderate AD showed greater, but not statistically significant, improvements in the ADAS-Cog compared to placebo. The fact that the subjects on placebo showed unexpected improvements may be a possible confound for the overall results.

The ongoing Phase II trials will expand the positive results reported from the initial Phase II study. This trial will assess cognition and global function in patients after 24 weeks treatment with SB-742457 (15 or 35 mg/kg) as compared to donepezil treatment or placebo, completed recruitment in summer

2010. This trial also includes secondary outcome measures of depressive symptoms as well as pharmacokinetics and exploratory pharmacogenetics (such as interactions with the APOEepsilon4 genotype) (Maher-Edwards, personal communication). A parallel trial, which also recently completed recruitment, is assessing SB-742457 at the same two doses given as adjunct therapy to patients already taking donepezil.

E. Lundbeck

Lundbeck is more aggressively pursuing 5-HT_6 antagonists for schizophrenia than dementia. Thus, there are no available data on clinical efficacy of its lead candidate antagonist, Lu AE58054. A clinical Phase II study consisting of 270 patients will investigate whether 24 weeks of Lu AE58054 treatment improves cognition and functional outcomes in patients with moderate Alzheimer's disease that are already in treatment with donepezil (ClinicalTrials.gov, 2010).

F. Memory Pharmaceuticals

MEM 68626, a 5-HT_6 antagonist, improved memory in F344 rats. Rowe *et al.* (2008) demonstrated improved performance in spatial memory using novel object recognition and radial maze. Memory Pharmaceuticals had nominated MEM 68626 as a lead development candidate for clinical trials; however, due to setbacks in funding this biopharmaceutical company (cofounded by Eric Kandel) merged with Roche in 2008. There is currently no public information available on development plans for MEM 68626.

G. Pfizer/Wyeth

SAM-531, a potent 5-HT_6 antagonist, is Pfizer/Wyeth's lead compound for Alzheimer's and related dementia. A Phase I trial assessing safety and tolerability of SAM-531 was successfully completed where subjects had no adverse effects and suitable plasma concentrations. Another Phase I study evaluated interactions between SAM-531 and gemfibrozil, a cytochrome P-450 2C8 inhibitor (clinicaltrials.gov, 2010). Gemfibrozil is used to treat high triglycerides and cholesterol. It is unclear what interactions are expected from this study, but it is possible that SAM-531 has unfavorable effects on blood lipids, or alternatively, gemfibrozil is acting as a bioavailability modulator. SAM-531 will also be assessed for interactions with verapamil in a Phase I trial, where verapamil is likely enhancing SAM-531brain bioavailability through inhibition of p-glycoprotein transporter efflux. Multiple doses of verapamil on the safety and plasma concentrations of a single dose of SAM-531 will be examined in healthy subjects.

Pfizer/Wyeth recently initiated a 52-week study of AD patients, which will test three doses (1.5, 3, or 5 mg) of SAM-531 or donepezil using the ADAS-Cog, Disability Assessment for Dementia (DAD), CANTAB, and the Alzheimer's Disease Cooperative Study-Clinical Global Impression of Change (ADCS-CGIC), with a planned finish in 2011.

The CANTAB is a computer-based neuropsychological test battery measuring aspects of vigilance, attention switching, working and long-term memory, and visual perception. For clinical trials tracking dementia, the CANTAB Paired Associate Learning (PAL), a visuospatial learning and episodic memory test, is often a primary outcome measure, due to its sensitivity in showing dementia progression (Yurko-Mauro *et al.*, 2010).

The ADCS-CGIC is a behavior and cognition assessment tool designed for clinic or home use flexibility for tracking patient progress. Because of its semiunstructured format, clinicians, subjects, or family members can focus on specific areas of dysfunction, which may make the ADCS-CGIS more sensitive than traditional AD measures.

Pfizer is also testing SAM-531's effects on physiological parameters of sleep quality in a separate study with healthy adults, using sleep electroencephalogram (EEG) and quantitative wake EEG (qEEG). Both SAM-531 (1, 3, 10 mg) and the active control, donepezil (5 mg), significantly increased sleep onset latency, latency to stage 1 sleep, latency to persistent sleep and percent time awake, or drowsy in the first time interval; thus indicating an alerting effect for both drugs. The QEEG results also supported this supposition. However, the main endpoint, duration of total REM sleep, was unchanged (Baird-Bellaire *et al.*, 2010).

H. Suven

Suven has done extensive preclinical research on its lead 5-HT_6 antagonists, SUVN-502 and SUVN-507. SUVN-502 increases acetylcholine and glutamate levels in the rat hippocampus and frontal cortex (Bhyrapuneni *et al.*, 2008). SUVN-502 has also been shown to enhance performance, as compared to vehicle, in the novel object recognition task and Morris water maze in young, unimpaired rats as well as those given scopolamine or MK-801 (Vishwakarma *et al.*, 2007).

Suven recently completed a Phase I clinical trial in Switzerland using a double-blind, placebo-controlled, randomized, ascending single-dose design in healthy males. The safety and tolerability of single ascending doses of SUVN-502 was found to be acceptable in terms of clinical chemistry, hematology, ECG, and urinalysis (Nirogi *et al.*, 2009). The pharmacokinetic profiles of single ascending doses of SUVN-502 and its active metabolite were shown to be appropriate for once-daily dosing and no adverse events were reported. A Phase II proof-of-concept

(POC) study is expected to start in 2010, although there has been some speculation that funds for such a trial have not been yet allocated.

I. Synosia

SYN-120, a potent and selective antagonist of the 5-HT_6 receptor, was discovered by Roche and is now under clinical development by Synosia. SYN-114 has completed a Phase I single ascending dose and multiple ascending dose study, and SYN-120 has completed a single ascending dose study (ClinicalTrials.gov). Both studies showed that the drugs were well tolerated and reported no adverse effects. The company plans to begin another Phase I trial of SYN-120 examining several doses and their effects on selected biomarkers in late 2010 (http://www.synosis.com/pipeline.html).

V. Clinical Studies with 5-HT_6 Agonists

A. 5-HT_6 Agonists

Due to the technical challenges of developing a specific, highly potent 5-HT_6 agonist, less is known about neurochemical and behavioral effects of 5-HT_6 stimulation. However, in the last 5 years several selective 5-HT_6 agonists have been identified and are now in clinical development.

The best studied 5-HT_6 agonist is WAY-181187, which, at doses of 3–30 mg/kg s.c., increases GABA release in the hippocampus, as shown via microdialysis, and attenuates stimulated glutamate (Schechter *et al.*, 2008). WAY-181187 had also been shown to increase GABA in frontal cortex (10–30 mg/kg) but decrease serotonin and dopamine release (30 mg/kg s.c.). Additionally, extracellular GABA is augmented in the striatum with systemic application of WAY-181187, but this same drug has no effect on GABA in the nucleus accumbens or thalamus. Such upregulation of inhibitory transmission in key learning regions suggests that 5-HT_6 agonists deleteriously affect hippocampal functioning. Indeed, WAY-181187 has been shown to attenuate hippocampal LTP over 100–300 nM (West *et al.*, 2009).

Several studies on 5-HT_6 agonists corroborate the expectation that 5-HT_6 activation is cognition-impairing. WAY-181187 disrupted Morris water maze performance in adult rats and impaired social recognition (i.p. or in cortex) while SB-271046 when applied directly to the cortex, but not striatum or nucleus basalis magnocellularis, reversed this effect (Loiseau *et al.*, 2008).

Interestingly, WAY-181187 had no effect on Morris water maze performance in old rats, suggesting that age-related changes in memory processing may be adaptive to the pharmacological effects of $5\text{-}HT_6$ stimulation. At any rate, this finding indicates that agonists will not be noticeably impairing in older brains. Novel object recognition (1 or 24 h intertrial interval or passive avoidance (1 or 24 h intertrial interval) performance in adult rats was also unchanged by WAY-181187 administration. WAY-181187 is currently being developed as an antidepressant (Carr *et al.*, 2010), which may prove to be useful in treating the BPSD in Alzheimer's patients.

There is a conference report and also two journal articles showing $5\text{-}HT_6$ agonist treatment as procognitive. The agonist, E-6801, and the inverse antagonist, E-6669, improved novel object recognition (4 h ITI) in adult rats (Pauwels *et al.*, 2006). In a more recent study, E-6801 (2.5 and 5 mg/kg) was compared to donepezil (0.03 and 1 mg/kg) and found to be as effective in reversing scopolamine deficits in a novel object recognition test (Kendall *et al.*, 2010). However, this was shown only at an ITI of 1 min and with a scopolamine dose of 0.5 mg/kg, a dose that may cause changes in motor activity. The authors proposed that $5\text{-}HT_6$ receptors on glutamatergic and cholinergic neurons were stimulating release of neurotransmitters and enhancing learning. This explanation does not account for the fact that a majority of $5\text{-}HT_6$ receptors appear to be on GABAergic neurons, or why other $5\text{-}HT_6$ agonists cause impairment in learning. There may be differences in potency of these agonists, where low activation has some learning enhancement but higher activation causes detriments.

B. Dimebolin

Dimebolin is antihistamine with diverse pharmacology for many receptors, one of them being a potent $5\text{-}HT_6$ inhibition. Dimebolin, currently prescribed in Russia for hay fever, demonstrated highly robust results in a Phase II-equivalent clinical trial in moderate AD patients. However, a more extensive clinical Phase II trial attempting to replicate these effects was recently reported to show no difference between dimebolin and placebo in AD subjects. Roughly 600 drug-free AD patients were given placebo, 15, or 60 mg dimebolin (licensed as Dimebon). Scores on the ADAS-Cog and CIBIC+ revealed no differences between groups. On the Neuropsychiatric Inventory (NPI), the Dimebon group showed a nonsignificant improvement over placebo but the ADCS-ADL (Alzheimer's Disease Cooperative Study-Activities of Daily Living) scores were similar across all groups.

Dimebolin has been described as a small-molecule mitochondrial permeability transition pore (MPTP) blocker, as well as an NMDA receptor antagonist and cholinesterase inhibitor. However, its action on other receptors and acetylcholinesterase are in the μM range, and thus not potent enough to explain the strong cognitive effects observed in the Russian trial. Yet, dimebolin was also reported to be a potent $5\text{-}HT_6$ antagonist in cAMP assays of recombinant receptors (50 nM).

Dimebolin has less affinity (10-fold less potent) for 5-HT$_6$ receptors than other antagonists such as SB-399885, but both improve social recognition at similar doses, which may be due to the complex pharmacological profile of dimebolin. In any case, dimebolin presents an example of possible increased efficacy due to multiple actions on key receptors and enzymes, and should be considered an example of how palliative monotherapies for dementia will undoubtedly fall by the wayside in favor of cocktails with diverse activities.

VI. Issues in Clinical Evaluation of 5-HT$_6$ Ligands

A. Ceiling Effects and Adjunct Therapy

Clinical studies of 5-HT$_6$ drugs have targeted patients with mild-to-moderate dementia. Subject selection of such patients is difficult, since the diagnosis of mild-to-moderate dementia uses tools that do not fully assess different types of cognitive ability. Therefore, the variability of cognitive ability in trial subjects can be quite high, and thus may obscure smaller scale benefits from a treatment. Additionally, the progress rate of disease can also be quite broad. The proposed palliative nature of 5-HT$_6$ antagonists means that the drugs will likely show weak effects in a subject whose disease progression is slow, since cognition needs to be decreasing at a fairly consistent rate in order to show a drug's protective efficacy. Not surprisingly pharmaceutical companies have undertaken to add extra testing time points and extra tests that measure more complex cognition than the MMSE, that has traditionally been used to assess a drug's efficacy on Alzheimer's and other related dementia.

Since 5-HT$_6$ ligands will not have a substantial effect on Alzheimer's pathology, they will likely be more efficacious during the same disease stages as acetylcholinesterase inhibitors, when the cholinergic system is still somewhat intact. However, as seen with animal models, young unimpaired rodents generally do not exhibit improvements in memory performance with 5-HT$_6$ antagonists. Such results imply that most 5-HT$_6$ ligands may not be as powerful as donepezil in modulating acetylcholine, since their action is indirect rather than direct. It may be that there is a narrow window when cholinergic function is compromised, but is not severely compromised in which 5-HT$_6$ antagonists may be expected to work. Several clinical studies are underway where subjects received 5-HT$_6$ antagonists as adjunct therapy to donepezil, operating on the theory that a differential mechanism of action on cholinergic function will allow synergistic enhancement. However, it is still not clear whether this combination will have additive effects on an already compromised supply of acetylcholine. Indeed, a recent study reported in abstract form suggests that some 5-HT$_6$ antagonists do not have significant effects on hippocampal or cortical acetylcholine levels as measured by microdialysis with SLV347, a 5-HT$_6$ antagonist, compared to donepezil (De Groote *et al.*, 2010). While donepezil (1 mg/kg p.o.) produced potent ACh increases in the hippocampus and prefrontal cortex, SLV347 (up to 30 mg/kg p.o.) had no effect on ACh, even though this dose reversed scopolamine deficits in a novel object recognition task (De Groote *et al.*, 2010). This finding suggests that, although 5-HT6 modulation of Ach has been touted as a primary mode of action, noncholinergic mechanisms may be key for the procognitive effects observed with 5-HT$_6$ blockade.

B. Peripheral Effects

Although there are few 5-HT$_6$ receptors in the periphery, side effects may still occur from metabolism and nonspecific action of 5-HT$_6$ ligands. Animal models such as rodents often have markedly different metabolic activities on certain groups of molecules than humans. A major concern is the extent of metabolite generation after a drug is administered, which often causes unforeseen side effects. Decreases in experimental drug blood concentration, as indicated by initial pharmacokinetic studies, can be controlled for by administering higher amounts. However, it is more difficult to predict if unique metabolites not seen in rodents will be formed in clinical bioavailability studies. The similar structure of some metabolites can sometimes cause them to behave in a similar manner, perhaps even synergizing with the original drug. Conversely, some uniquely human metabolites may have the opposite effect on a target as the parent drug, canceling out all therapeutic action. Thus far, no Phase I studies using 5-HT$_6$ ligands have reported unanticipated metabolites or unacceptable side effects. The main side effect of 5-HT$_6$ ligands appear to be liver toxicity, which was initially observed in rodent testing and thus precluded some ligands from clinical development such as RO 4368554 (R. Schrieber, personal communication).

C. Bioavailability

Initial evaluation of 5-HT$_6$ antagonists on various aspects of cognition produced mixed results (Mitchell and Neumaier, 2005; Rogers and Hagan, 2001; Woolley *et al.*, 2003), so much so that there was substantial reason to believe that 5-HT$_6$ receptors had insignificant roles in learning. Further examination of bioavailability of RO 046790 raised doubts on the brain penetration of the ligand; however, this was only demonstrated conclusively after it was tested in parallel with newer ligands. Indeed, RO 046790 and RO 657199, compounds found to be inferior to the more bioavailable SB-270146, had no effect on latent inhibition (Leng *et al.*, 2003). Later it was shown that many of the earliest 5-HT_6 compounds had low brain penetration, which accounted for the conflicting reports of 5-HT_6 receptor blockade on animal cognition. This problem of bioavailability is still a considerable issue, since several drugs developed as Alzheimer's therapy, which achieved promising results in animal studies, have also been shown to have low brain penetration when administered to humans. This was the case for SB-271046, whose brain to blood ratio in humans was rather poor, as compared to ratios seen in rodents (0.5:1). SB-271046 was the first 5-HT_6 antagonist to be tested in Phase I trials; however, this trial had to be halted due to the unexpected lack of blood–brain barrier permeability (Garfield and Heisler, 2009).

D. Palliative Versus Curative Properties of 5-HT_6 Ligands

Due to the strong evidence of cholinergic action, 5-HT_6 antagonists and agonists were considered to have primarily palliative potential. Manipulation of 5-HT_6 receptors, as with many other Gs-protein-coupled receptors, was hypothesized to have little influence on the pathological biomarkers of Alzheimer's: buildup of beta-amyloid and tau deposition. Yet it still not clear whether amyloid is the primary cause of the disease or simply a consequence of a more global mechanism of dementia such as nonspecific protein deposition in neurons, decreased regenerative capability, or immune system dysfunction. However, there are some small findings that point to curative properties of 5-HT_6 ligands. The strongest indications of long-term palliative action come from the findings of 5-HT_6 receptor modulation of signaling factors controlling growth and energy use such as mTOR and Jab1. Additionally, some animal behavior data suggest that chronic 5-HT_6 blockade may improve cognitive function even after the drug has washed out (Mitchell *et al.*, 2009). However, it must be noted that any potential curative 5-HT_6 effects are likely to be quite subtle compared to those needed to effectively combat Alzheimer's pathological mechanisms. Indeed, all treatments currently in development thus far may not be powerful or safe enough to reverse the heterogeneous and complex pathology of dementia. Thus it is within the best interest of all those at risk for dementia (i.e., everyone) that many types of treatments, even those with subtle efficacy, become available for dementia treatment.

VII. Conclusion

Many promising 5-HT_6 ligands are currently being investigated either preclinically or via human trials. However, with the current constraints on drug development, including funding, lack of sensitive evaluation tools, and increasing bureaucracy in all stages of research, it is uncertain whether any of these drugs will reach the market. 5-HT$_6$ ligands have many potential applications, improving not just long-term memory, but also attention, working memory, and behavioral issues such as depression (BPSD). Thus far most 5-HT$_6$ ligands completing Phase I clinical trials have been well tolerated with minimal side effects. What remains to be seen is the extent of their clinical efficacy in ameliorating dementia progression. Pharmaceutical companies are now exploring 5-HT$_6$ ligands as adjunct therapy. With further mechanistic understanding of 5-HT$_6$-mediated cellular signaling via preclinical research, clinical studies can be designed with more sensitive tests, appropriate endpoints, and patient populations; thus, 5-HT$_6$ drugs may soon be available to patients in the coming years.

References

Ahlander, M., Misane, I., Schött, P.A. and Ogren, S.O. (1999). A behavioral analysis of the spatial learning deficit induced by the NMDA receptor antagonist MK-801 (dizocilpine) in the rat. *Neuropsychopharmacology* **21**(3); 414–426.

Baird-Bellaire, S. J., Wan, H., Vandenhende, F., Plotka, A., Poola, N., Paul, J., *et al.* (2010). The effects of a single dose administration of SAM-531, a 5-HT$_6$ antagonist, on sleep electroencephalogram (EEG) and quantitative Wake EEG (qEEG) in healthy subjects. International Conference on Alzheimer's and Parkinson's Diseases (ICAD), 203.

Bhyrapuneni, G., Muddana, N., Benade, V., Saralaya, S., Ajjala, D. and Mudigonda, K *et al.* (2008). The 5-HT$_6$ receptor antagonist SUVN-502 enhances acetylcholine and glutamate in rat ventral hippocampus and frontal cortex: a microdialysis study. *Alzheimer's Dementia* **4**(Suppl. 4); T495.

Bibancos, T., Jardim, D.L., Aneas, I. and Chiavegatto, S. (2007). Social isolation and expression of serotonergic neurotransmission-related genes in several brain areas of male mice. *Genes Brain Behav.* **6**(2); 529–539.

Blundell, J., Kouser, M. and Powell, C.M. (2008). Systemic inhibition of mammalian target of rapamycin inhibits fear memory reconsolidation. *Neurobiol. Learn. Mem.* **90**(1); 28–35.

Bourson, A., Borroni, E., Austin, R.H., Monsma, F.J. and Sleight, A.J. (1995). Determination of the role of the 5-ht$_6$ receptor in the rat brain: a study using antisense oligonucleotides. *J. Pharmacol. Exp. Ther.* **274**(1); 173–180.

Braithwaite, S.P., Paul, S., Nairn, A.C. and Lombroso, P.J. (2006). Synaptic plasticity: one STEP at a time. *Trends Neurosci.* **29**, 452–458.

Callaghan, C. K., Wan, B., Wang, W., Maloney, M., Milton, H., Della Chiesa, A., *et al.* (2009). 5HT$_6$ receptor antagonist SB742457 recovers an age related deficit in working memory but not in spatial memory in the rat. SfN Neuroscience Meeting Planner, 535.15.

Carr, G. V., Schechter, L. E., and Lucki, I. (2010). Antidepressant and anxiolytic effects of selective 5-HT$_{(6)}$ receptor agonists in rats. *Psychopharmacology*. DOI: 10.1007/s00213-010-1798-7.

Chin, J., Palop, J.J., Puolivali, J., Massaro, C., Bien-Ly, N. and Gerstein, H *et al.* (2005). Fyn kinase induces synaptic and cognitive impairments in a transgenic mouse model of Alzheimer's disease. *J. Neurosci.* **25**, 9694–9703.

Clinical Trials Database. (2003). Available at <http://clinicaltrials.gov/ct2/show/NCT01019421?term=lu+ae&rank=2>, accessed 07/07/10.

Costa, V.D., Duchatelle, P., Boulouard, M. and Dauphin, F. (2009). Selective 5-HT$_6$ receptor blockade improves spatial recognition memory and reverses age-related deficits in spatial recognition memory in the mouse. *Neuropsychopharmacology* **34**, 488–500.

Costa-Aze, V.D., Quiedeville, A., Anfray, J., Dauphin, F. and Boulouard, M. (2009). Effects of the blockade of the 5-HT$_6$ receptors on scopolamine-induced deficits of working memory and episodic-like memory in the mouse. *Fund. Clin. Pharmacol.* **23**, 350.

De Foubert, G., O'Neill, M.J. and Zetterstrom, T.S. (2007). Acute onset by 5-HT$_6$-receptor activation on rat brain brain-derived neurotrophic factor and activity-regulated cytoskeletal-associated protein mRNA expression. *Neuroscience* **147**, 778–785.

De Groote, L., Prickaerts, J., Reneerkens, O., Van Loevezijn, A. and De Bruin, N. (2010). The 5-HT$_6$ receptor antagonist SLV347 improves cognitive deficits associated with the cholinergic system. *FENS Abstr.* **5**, 057.18.

de Jongh, R., Bolt, I., Schermer, M. and Olivier, B. (2008). Botox for the brain: enhancement of cognition, mood and pro-social behavior and blunting of unwanted memories. *Neurosci. Biobehav. Rev.* **32**(4); 760–776.

Foley, A.G., Hirst, W.D., Gallagher, H.C., Barry, C., Hagan, J.J. and Upton, N *et al.* (2008). The selective 5-HT$_6$ receptor antagonists SB-271046 and SB-399885 potentiate NCAM PSA immunolabeling not neurogenesis, in the hippocampal of dentate granule cells, but formation of mature Wistar rats. *Neuropharmacology* **54**, 1166–1174.

Foley, A.G., Murphy, K.J., Hirst, W.D., Gallagher, H.C., Hagan, J.J. and Upton, N *et al.* (2004). The 5-HT$_6$ receptor antagonist SB-271046 reverses scopolamine-disrupted consolidation of a passive avoidance task and ameliorates spatial task deficits in aged rats. *Neuropsychopharmacology* **29**, 93–100.

Garcia-Alloza, M., Hirst, W.D., Chen, C.P.L.H., Lasheras, B., Francis, P.T. and Ramirez, M.J. (2004). Differential involvement of 5-HT1B/1D and 5-HT$_6$ receptors in cognitive and non-cognitive symptoms in Alzheimer's disease. *Neuropsychopharmacology* **29**, 410–416.

Garfield, A.S. and Heisler, L.K. (2009). Pharmacological targeting of the serotonergic system for the treatment of obesity. *J. Physiol.* **15**(587); 49–60.

Geldenhuys, W.J. and Van der Schyf, C.J. (2009). The serotonin 5-HT$_6$ receptor: a viable drug target for treating cognitive deficits in Alzheimer's disease. *Expert. Rev. Neurother.* **9**, 1073–1085.

Halford, J.C.G., Harrold, J.A., Boyland, E.J., Lawton, C.L. and Blundell, J.E. (2007). Serotonergic drugs – effects on appetite expression and use for the treatment of obesity. *Drugs* **67**, 27–55.

Heal, D.J., Smith, S.L., Fisas, A., Codony, X. and Buschmann, H. (2008). Selective 5-HT$_6$ receptor ligands: progress in the development of a novel pharmacological approach to the treatment of obesity and related metabolic disorders. *Pharmacol. Ther.* **117**(2); 207–231.

Hirst, W.D., Abrahamsen, B., Blaney, F.E., Calver, A.R., Aloj, L. and Price, G.W *et al.* (2003). Differences in the central nervous system distribution and pharmacology of the mouse 5-hydroxytryptamine-6 receptor compared with rat and human receptors investigated by radioligand binding, site-directed mutagenesis, and molecular modeling. *Mol. Pharmacol.* **64** (6); 1295–1308.

Hirst, W.D., Stean, T.O., Rogers, D.C., Sunter, D., Pugh, P. and Moss, S.F *et al.* (2006). SB-399885 is a potent, selective 5-HT$_6$ receptor antagonist with cognitive enhancing properties in aged rat water maze and novel object recognition models. *Eur. J. Pharmacol.* **553**, 109–119.

Hoeffer, C.A. and Klann, E. (2010). mTOR signaling: at the crossroads of plasticity, memory and disease. *Trends Neurosci.* **33**(2); 67–75.

Huerta-Rivas, A., Perez-Garcia, G., Gonzalez-Espinosa, C. and Meneses, A. (2010). Time-course of 5-HT$_6$ receptor mRNA expression during memory consolidation and amnesia. *Neurobiol. Learn. Mem.* **93**, 99–110.

Isosaka, T., Hattori, K., Kida, S., Kohno, T., Nakazawa, T. and Yamamoto, T *et al.* (2008). Activation of Fyn tyrosine kinase in the mouse dorsal hippocampus is essential for contextual fear conditioning. *Eur. J. Neurosci.* **28**, 973–981.

Kay, G. (2004). The effect of adderall XR treatment on driving performance in young adults with ADHD. USPMHC Meeting Abstracts, 316.

Kelland, M.D., Soltis, R.P., Boldry, R.C. and Walters, J.R. (1993). Behavioral and electrophysiological comparison of ketamine with dizocilpine in the rat. *Physiol. Behav.* **54**, 547–554.

Kendall, I., Slotten, H.A., Codony, X., Burgueno, J., Pauwels, P.J., Vela, J.M. and Fone, K.C. (2010). E-6801, a 5-$HT_{(6)}$ receptor agonist, improves recognition memory by combined modulation of cholinergic and glutamatergic neurotransmission in the rat. *Psychopharmacology (Berl).*

King, M.V., Fone, K.C.F., Shacham, S., Fichman, M., Melendez, R. and Orbach, P *et al.* (2006a). Combination of sub-threshold doses of Aricept and PRX-07034, a novel 5-HT_6 receptor antagonist, enhances Novel Object Discrimination (NOD) memory. *J. Pharmacol. Sci.* **101**, 150.

King, M.V., Sleight, A.J., Woolley, M.L., Topham, I.A., Marsden, C.A. and Fone, K.C.F. (2004). 5-HT_6 receptor antagonists reverse delay-dependent deficits in novel object discrimination by enhancing consolidation – an effect sensitive to NMDA receptor antagonism. *Neuropharmacology* **47**, 195–204.

King, M.V., Spicer, C.H., Sleight, A.J., Fone, K.C.F. and Marsden, C.A. (2006b). Influence of median or dorsal raphe 5,7-DHT lesions on 5-ht_6 receptor antagonist activity in a Novel Object Discrimination paradigm. *J. Pharmacol. Sci.* **101**, 125.

Leng, A., Ouagazzal, A., Feldon, J. and Higgins, G.A. (2003). Effect of the 5-HT_6 receptor antagonists Ro 04-6790 and Ro 65-7199 on latent inhibition and prepulse inhibition in the rat: comparison to clozapine. *Pharmacol. Biochem. Behav.* **75**, 281–288.

Lim, S.H., Kwon, S.K., Lee, M.K., Moon, J., Jeong, D.G. and Park, E *et al.* (2009). Synapse formation is regulated by protein tyrosine phosphatase receptor T through interaction with cell adhesion molecules and Fyn. *EMBO J.* **28**, 3564–3578.

Lindner, M.D., Hodges Jr., D.B., Hogan, J.B., Orie, A.F., Corsa, J.A. and Barten, D.M *et al.* (2003). An assessment of the effects of serotonin 6 (5-HT_6) receptor antagonists in rodent models of learning. *J. Pharmacol. Exp. Ther.* **307**(2); 682–691.

Liy-Salmeron, G. and Meneses, A. (2007). Role of 5-HT1-7 receptors in short- and long-term memory for an autoshaping task: intrahippocampal manipulations. *Brain Res.* **1147**, 140–147.

Liy-Salmeron, G. and Meneses, A. (2008). Effects of 5-HT drugs in prefrontal cortex during memory formation and the ketamine amnesia-model. *Hippocampus* **18**, 965–974.

Lungu, G.F., Stoica, G. and Wong, P.K. (2008). Down-regulation of Jab1, HIF-1alpha, and VEGF by Moloney murine leukemia virus-ts1 infection: a possible cause of neurodegeneration. *J. Neurovirol.* **14**(3); 239–251.

Maher-Edwards, G., Zvartau-Hind, M., Hunter, A.J., Gold, M., Hopton, G., Jacobs, G., Davy M., *et al.* (2009). Double-blind, controlled Phase II study of a 5-ht_6 receptor antagonist, SB-742457, in Alzheimer's disease. *Curr. Alzheimer Res.* 7, 374–85.

Marcos, B., Cabero, M., Solas, M., Aisa, B. and Ramirez, M.J. (2009). Signalling pathways associated with 5-HT_6 receptors: relevance for cognitive effects. *Int. J. Neuropsychopharmacol.* **9**, 1–10.

Marcos, B., Chuang, T.T., Gil-Bea, F.J. and Ramirez, M.J. (2008). Effects of 5-HT_6 receptor antagonism and cholinesterase inhibition in models of cognitive impairment in the rat. *Br. J. Pharmacol.* **155**, 434–440.

Meffre, J., Dupuis, D., Seveno, M., Mannoury, M., La Cour, C., Dekeyne A., *et al.* (2010). Proteomic and functional analysis suggests that 5-HT_6 receptors compromise cognition by recruiting mTOR pathway. *FENS Abstr.* 5, 077.8.

Meneses, A. (2001a). Effects of the 5-HT_6 receptor antagonist Ro 04-6790 on learning consolidation. *Behav. Brain Res.* **118**, 107–110.

Meneses, A. (2001b). Role of 5-HT_6 receptors in memory formation. *Drug News Perspect.* **14**, 396–400.

Mitchell, E.S., Hoplight, B.J., Lear, S.P. and Neumaier, J.F. (2006). BGC20-761, a novel tryptamine analog, enhances memory consolidation and reverses scopolamine-induced memory deficit in social and visuospatial memory tasks through a 5-HT_6 receptor-mediated mechanism. *Neuropharmacology* **50**, 412–420.

Mitchell, E.S., McDevitt, R.A. and Neumaier, J.F. (2009). Adaptations in 5-HT receptor expression and function: implications for treatment of cognitive impairment in aging. *J. Neurosci. Res.* **87**, 2803–2811.

Mitchell, E.S. and Neumaier, J.F. (2005). 5-HT_6 receptors: a novel target for cognitive enhancement. *Pharmacol. Ther.* **108**, 320–333.

Mitchell, E.S. and Neumaier, J.F. (2008). 5-HT_6 receptor antagonist reversal of emotional learning and prepulse inhibition deficits induced by apomorphine or scopolamine. *Pharmacol. Biochem. Behav.* **88**, 291–298.

Morairty, S.R., Hedley, L., Flores, J., Martin, R. and Kilduff, T.S. (2008). Selective 5HT(2A) and $5HT_{(6)}$ receptor antagonists promote sleep in rats. *Sleep* **31**, 34–44.

Nelson, D.L. (2005). The 5-HT_6 receptor: filled with promise or promises unfilled? *Neuropsychopharmacology* **30**, S53.

Nirogi, R., Kambhampati, R. and Shinde, A *et al.* (2009). SUVN-502: a safe, potent, brain penetrant and orally active 5-HT_6 receptor antagonist – first in human clinical phase-1 study. *Alzheimer's Dementia* **5**(Suppl. 1); P1–251.

Pauwels, P., Slotten, H., Fone, K., Merce, R., Buschmann, H. and Codony, X. (2006). Memory enhancing properties of E-6801, a 5-HT_6 receptor ligand with agonist properties, in the novel object discrimination test. *Acta Pharmacol. Sin.* **27**, 76.

Pena, F., Ordaz, B., Balleza-Tapia, H., Bernal-Pedraza, R., Marquez-Ramos, A. and Carmona-Aparicio, L *et al.* (2010). Beta-amyloid protein (25–35) disrupts hippocampal network activity: role of Fyn-kinase. *Hippocampus* **20**, 78–96.

Pitsikas, N., Zisopouiou, S., Pappas, I. and Sakellaridis, N. (2008). The selective 5-HT_6 receptor antagonist Ro 04-6790 attenuates psychotomimetic effects of the NMDA receptor antagonist MK-801. *Behav. Brain Res.* **188**, 304–309.

Roberts, J.C., Reavill, C., East, S.Z., Harrison, P.J., Patel, S. and Routledge, C *et al.* (2002). The distribution of 5-HT_6 receptors in rat brain: an autoradiographic binding study using the radiolabelled 5-HT_6 receptor antagonist [I-125]SB-258585. *Brain Res.* **934**, 49–57.

Robillard, J. and Christie, B. (2007). Role of 5-HT_6 receptors in hippocampal bidirectional synaptic plasticity. *Can. J. Exp. Psychol.* **61**, 357.

Rogers, D.C. and Hagan, J.J. (2001). 5-HT_6 receptor antagonists enhance retention of a water maze task in the rat. *Psychopharmacology* **158**, 114–119.

Rowe, W., Lowe, D., *et al.* (2008). Characterization of serotonin 5-HT_6 receptor antagonists as putative drugs for age-related mild cognitive impairment and Alzheimer's disease. SfN Neuroscience Meeting Planner, 685.9.

Sambeth, A., Riedel, W.J., Smits, L.T. and Blokland, A. (2007). Cholinergic drugs affect novel object recognition in rats: relation with hippocampal EEG? *Eur. J. Pharmacol.* **572**, 151–159.

Schechter, L.E., Lin, Q., Smith, D.L., Zhang, G.M., Shan, Q. and Platt, B *et al.* (2008). Neuropharmacological profile of novel and selective 5-HT_6 receptor agonists: WAY-181187 and WAY-208466. *Neuropsychopharmacology* **33**, 1323–1335.

Tkachenko, S., Ivachtchenko, A., Khvat, A., Okun, I., Lavrovsky, Y., and Salimov, R. (2009). Discovery and preclinical studies of AN-322, a highly selective and potent 5-HT_6 antagonist for cognition enhancement in treating neurodegenerative diseases. International Conference on Alzheimer's and Parkinson's Diseases (ICAD), 490.

Upton, N., Chuang, T.T., Hunter, A.J. and Virley, D.J. (2008). 5-HT_6 receptor antagonists as novel cognitive enhancing agents for Alzheimer's disease. *Neurotherapeutics* **5**, 458–469.

Vishwakarma, S., Benade, V., and Marshal, V. (2007). SUVN-502: a potent 5-HT_6 receptor antagonist reverses MK-801 induced amnesia and enhances brain glutamate levels. SfN Neuroscience Meeting Planner, 745.5/CCC7.

West, P.J., Marcy, V.R., Marino, M.J. and Schaffhauser, H. (2009). Activation of the 5-HT_6 receptor attenuates long-term potentiation and facilitates GABAergic neurotransmission in rat hippocampus. *Neuroscience* **164**, 692–701.

Woolley, M.L., Marsden, C.A., Sleight, A.J. and Fone, K.C.F. (2003). Reversal of a cholinergic-induced deficit in a rodent model of recognition memory by the selective 5-HT_6 receptor antagonist, Ro 04-6790. *Psychopharmacology* **170**, 358–367.

Xu, F., Plummer, M.R., Len, G.W., Nakazawa, T., Yamamoto, T., Black, I.B. and Wu, K. (2006). Brain-derived neurotrophic factor rapidly increases NMDA receptor channel activity through Fyn-mediated phosphorylation. *Brain Res.* **1121**, 22–34.

Yun, H.M., Baik, J.H., Kang, I., Jin, C. and Rhim, H. (2010). Physical interaction of Jab1 with human serotonin 6 G-protein-coupled receptor and their possible roles in cell survival. *J. Biol. Chem.* **285**, 10016–10029.

Yun, H.M., Kim, S., Kim, H.J., Kostenis, E., Kim, J.I. and Seong, J.Y *et al.* (2007). The novel cellular mechanism of human 5-HT_6 receptor through an interaction with Fyn. *J. Biol. Chem.* **282**, 5496–5505.

Yurko-Mauro, K., McCarthy, D., Rom, D., Nelson, E. B., Ryan, A. S., Blackwell, A., *et al.* (2010). Beneficial effects of docosahexaenoic acid on cognition in age-related cognitive decline. *Alzheimer's Dementia.* 6, 456–64.

OTHER 5-HT$_6$ RECEPTOR-MEDIATED EFFECTS

Franco Borsini

Sigma-Tau Industrie Farmaceutiche, Riunite S.P.A., Pomezia, Italy

The role of 5-HT$_6$ receptors has been investigated for the treatment of depression, anxiety, schizophrenia, cognitive disorders, obesity, and addiction. Moreover, there is evidence that 5-HT$_6$ receptors may also play a role in animal models of Parkinson's disease (PD), convulsive states, inflammation, and sleep.

I. Animal Paradigms to Test Potential Antiparkinson's Disease Drugs

PD is a chronic, progressive neurodegenerative disorder, characterized by a gradual loss of dopaminergic neurons in the pars compact of the substantia nigra and subsequent depletion of dopamine (DA) in the striatum. Due to the fact that 5-HT$_6$ receptors may differently influence DA in the various brain areas (Dawson, 2011; Di Chiara *et al.*, 2011), only experiments dealing with striatal DA are considered in this paragraph.

One of the behavioral paradigms to measure drug activity in PD is the behavior that 6-OHDA-treated rats display after unilateral 6-OHDA injection into medial forebrain bundle, which contains dopaminergic fibers that impinge striatum. 6-OHDA is a neurotoxin that destroys dopaminergic fibers. After an appropriate period of time, when striatal postsynaptic dopaminergic receptors become hypersensitive, the administration of the antiparkinson drug benserazide + levodopa induces contralateral rotations. Muscarinic antagonists and amphetamine, in contrast, induce ipsilateral rotations. In such rat animal model, intraperitoneal administrations of the 5-HT$_6$ receptor antagonist Ro 04-6790 reduced atropine- and scopolamine-induced ipsilateral rotations (Bourson *et al.*, 1998) but neither modulated amphetamine-induced ipsilateral or levodopa-induced contralateral rotations (Bourson *et al.*, 1998) nor induced contralateral or ipsilateral turnings *per se* (Bourson *et al.*, 1998).

INTERNATIONAL REVIEW OF NEUROBIOLOGY, VOL. 96
DOI: 10.1016/B978-0-12-385902-0.00008-5

Another model used as an animal paradigm of PD is haloperidol-induced catalepsy, whose principle is based on dopaminergic receptor blockade in basal ganglia. In contrast to levodopa, Ro 04-6790 did not modulate haloperidol-induced catalepsy in mice (Bourson *et al.*, 1998). However, Ro 04-6790 does not bind to mouse 5-HT_6 receptors (Hirst *et al.*, 2003), which may well account for its inefficacy in mice.

When the 5-HT_6 receptor antagonist SB-271046 was administered subcutaneously in rats, the neurochemical effects of amphetamine on striatal DA were increased (Dawson *et al.*, 2003). In the same experiment, it was shown that SB-271046 did not change DA levels in the striatum.

Thus, the role of 5-HT_6 receptor antagonists is not well definite and understood as possible involvement in PD.

II. Miscellanea

The 5-HT_6 receptor antagonists SB-271046, SB-258510, and Ro 04-6790 (Routledge *et al.*, 2000), SB-399885 (Hirst *et al.*, 2006), and SB-357134 (Stean *et al.*, 2002) were reported to exert anticonvulsant effects against electroshock-induced seizures.

The 5-HT_6 receptor antagonists, Ro4368554 was reported to promote sleep (Morairty *et al.*, 2008), and SB-258585 and SB-399885 to exert anti-inflammatory effects (Castaneda-Corral *et al.*, 2009).

References

Bourson, A., Boess, F.G., Bos, M., and Sleight, A.J. (1998). Involvement of 5-HT_6 receptors in nigrostriatal function in rodents. *Br. J. Pharmacol.* **125**, 1562–1566.

Castaneda-Corral, G., Rocha-Gonzales, H.I., Araiza-Saldana, C.I., Ambriz-Tututi, M., Vidal-Cantu, G.C., and Granados-Soto, V. (2009). Role of peripheral and spinal 5-HT_6 receptors according to the rat formalin test. *Neuroscience* **162**, 444–452.

Dawson, L. (2011). The central role of 5-HT_6 receptors in modulating brain neurochemistry. In: Pharmacology of 5-HT_6 Receptors, Part II (F. Borsini, ed.), International Review of Neurology, Vol. 96, Elsevier, Amsterdam.

Dawson, L.A., Nguyen, H.Q., and Li, P. (2003). Potentiation of amphetamine-induced changes in dopamine and 5-HT by a 5-HT_6 receptor antagonist. *Brain Res. Bull.* **59**, 513–521.

Di Chiara, G., Valentini, V., and Fenu, S. (2011). Behavioral and neurochemical pharmacology of 5-HT_6 receptors related to reward and reinforcement. In: Pharmacology of 5-HT_6 Receptors, Part II (F. Borsini, ed.), International Review of Neurology, Vol. 96, Elsevier, Amsterdam.

Hirst, W.D., Abrahamsen, B., Blaney, F.E., Calver, A.R., Aloj, L., Price, G.W., and Medhurst, A.D. (2003). Differences in the central nervous system distribution and pharmacology of the mouse 5-hydroxytryptamine-6 receptor compared with rat and human receptors investigated by radioligand binding, site-directed mutagenesis, and molecular modelling. *Mol. Pharmacol.* **64**, 1295–1308.

Hirst, W.D., Stean, T.O., Rogers, D.C., Sunter, D., Pugh, P., Moss, S.F., Bromidge, S.M., Riley, G., Smith, D.R., Bartlett, S., Heidbreder, C.A., Atkins, A.R., Lacroix, L.P., Dawson, L.A., Foley, A.G., Regan, C.M., and Upton, N. (2006). SB-399885 is a potent, selective 5-HT$_6$ receptor antagonist with cognitive enhancing properties in aged rat water maze and novel recognition models. *Eur. J. Pharmacol.* **553**, 109–119.

Morairty, S.R., Hedley, L., Flores,J., Martin, R., and Kilduff, T.S. (2008). Selective 5-HT$_{2A}$ and 5-HT$_6$ receptor antagonists promote sleep in rats. *Sleep* **31**, 34–44.

Routledge, C., Bromidge, S.M., Moss, S.F., Price, G.W., Hirst, W., Newmann, H., Riley, G., Gager, T., Stean, T., Upton, U., Clarke, S.E., Brown, A.M., and Middlemiss, D.N. (2000). Characterization of SB-271046: a potent, selective and orally active 5-HT$_6$ receptor antagonist. *Br. J. Pharmacol.* **130**, 1606–1612.

Stean, T.O., Hirst, W.D., Thomas, D.R., Price, G.W., Rogers, D., Riley, G., Bromidge, S.M., Serafiniwska, H.T., Smith, D.R., Bartlett, S., Deeks, N., Duxon, M., and Upton, N. (2002). Pharmacological profile of SB-357134: a potent, selective, brain penetrant, and orally active 5-HT$_6$ receptor antagonists. *Pharmacol. Biochem. Behav.* **71**, 645–654.

Index

I

J

K

L

M

N

O

P

Q

R

CONTENTS OF RECENT VOLUMES

Volume 39

Volume 40

Volume 41

Volume 42

Volume 43

Volume 44

Volume 45

Volume 46

Volume 47

Volume 48

Volume 49

Volume 50

Volume 51

Volume 52

Volume 53

Volume 54

Volume 55

Volume 56

Volume 57

Volume 58

Volume 59

Volume 60

Volume 61

Volume 62

Volume 63

Volume 64

Volume 65

Volume 66

Volume 67

Volume 68

Volume 69

Volume 70

Volume 71

Volume 72

Volume 73

Volume 74

Volume 75

Volume 76

Volume 77

Volume 78

Volume 79

Volume 80

Volume 81

Volume 82

Volume 83

Volume 84

Volume 85

Volume 86

Volume 87

Volume 88

Volume 89

Volume 90

Volume 91

Volume 92

Volume 93

Volume 94

Volume 95

Edwards Brothers Inc.
Ann Arbor MI. USA
April 11, 2011